# UNDERSTANDING THE TWO STATE UNIVERSE

by
Angelo A. Molinaro

INFINITY
PUBLISHING

*ISBN 978-0-7414-0570-8*

INFINITY PUBLISHING
1094 New DeHaven Street, Suite 100
West Conshohocken, PA 19428-2713
Toll-free (877) BUY BOOK
Local Phone (610) 941-9999
Fax (610) 941-9959
Info@buybooksontheweb.com
www.buybooksontheweb.com

To Mr. and Mrs. Anthony Molinaro
my father and mother

# PREFACE

The purpose of this book is to describe, simplify and make understandable, in an interesting way; the Two State Universe Theory that was presented in my previous book titled The Two State Universe.

Since my previous book described a new and different theory of the universe, it required a technical explanation, the coining of new words and a considerable amount of mathematics to describe the new theory and its applications.

In this book we will use little or no mathematics and keep the technical explanations as simple and as interesting as possible, in order that the general reader and students having an interest in science will be able to understand the Two State Universe Theory and its applications.

Angelo A. Molinaro

# CONTENTS

# APPENDIX

# INTRODUCTION

By October 9, 1983 science had progressed to a high degree of sophistication. It was so sophisticated that it enabled us to go to the moon and back. It also showed us how to split the atom and use the vast amount of energy in the nucleus, of the atom, for peaceful purposes as well as for destructive weapons of war. Progress was being made in all areas of science, but in spite of these great achievements there were cracks in the various methods and theories of science that were in use at that time.

In spite of all the knowledge accumulated and the vast amount of techniques available to solve problems, there were many unsolved problems that puzzled the scientists. Also, there were conflicts between the different theories and ideas that were known at that time. For example, there were conflicts between Newton's theories and Einstein's theories, and there were conflicts between classical physics and quantum mechanics. Newton's theory of motion assumes that mass is a constant. Einstein's theory assumes that mass increases with velocity. Newton's theory of gravity assumes that mass is attracted to mass. Einstein's theory of gravity assumes that mass warps space-time and the warped space-time is what causes masses to attract each other. Classical physics assumes that energy is continuous; quantum mechanics assumes that energy is not continuous, rather it travels in discrete bundles, and it is discontinuous.

There were also numerous paradoxes and riddles that puzzled the scientists. For example, scientists did not know what light was. There were two theories about light. Newton's theory which assumed that light was a particle, and Huygen's theory which assumed that light was a wave. Scientists are still trying to determine whether light is a wave or a particle. Any experiment that the scientists carried out showed that light was either a particle or a wave. One type of experiment would show the particle but not the wave. Another experiment would show the wave but not the particle. Therefore, depending upon the experiment performed either the particle nature of light would be detected or the wave nature of light would be detected. The scientists were unable to show both the wave and the particle nature of light in the same experiment.

Many other riddles of science appeared to have no solutions; such as, the riddle of the irresistible force and the immovable object. Science had no means to determine what happens when an irresistible force meets an immovable object. All the known theories at that time were unable to solve this problem. Another riddle that appeared to have no solution and could not be solved with any of the known theories was the riddle of the chicken or the egg. The known theories didn't have any means for solving this paradoxical problem of what came first, the chicken or the egg. There was also the riddle of what happens if you continue to go straight up in space. Does the universe end? Or does it continue to go straight up forever? The known theories could not answer any of these questions, because they didn't have the laws, formulas or means needed for solving any of these riddles or paradoxical problems.

Each new theory formulated appeared to answer most of the unknown problems asked by the scientists at the time the theory was formulated. However, later other problems would occur that could not be handled by the theory. The theory was then altered or replaced by another theory. Each new theory appeared to be logical and capable of answering all of the problems being asked at that time.

There were times when the scientists believed that no further theories had to be derived, because all of the questions of physics could be answered by the present knowledge. As soon as this belief became generally accepted some new crack in the known theories would appear. Scientists were accumulating knowledge at an accelerated rate, resulting in their theories becoming more sophisticated and more complicated. When some of these popular theories began to show cracks, the scientists would try to save them with wild imaginative ideas, which even further complicated the theories.

Einstein, however, like many scientists, believed that science should be simple and beautiful. He worked for many years until his death in I955 trying to find a unified field theory that would combine all the known forces of nature in one equation. At that time, the electrostatic force and the electromagnetic force yielded to a single equation. However, the force of gravity did not seem to fit in with these two forces.

Scientists are presently spending a considerable amount of time, effort and money looking for the unified field theory which will include all of the known forces of nature in one equation, and hopefully this unified field theory equation could be used to solve any scientific problem. This is a small sample of how science was progressing by October 9, I983, and some of the cracks in the theories that I was aware of at that time.

Then, on Saturday October 8, my brother asked me to take a trip to Pennsylvania. He owned a house in Forest City, a small town located near Scranton and Carbondale. We were both born in Carbondale, and we would take frequent trips to Pennsylvania, sometimes for a vacation and sometimes to do work on his house.

During most of the trips, my brother, my cousin, who lived near Scranton, and I would get together and have a rap session until the wee hours of the next morning. We would talk about everything, from patents and patent law to science and theories of the universe. During our discussion, on the night of October 8, we talked about all the apparent conflicts of science and many of the paradoxes that existed which appeared to have no solutions using the known theories of physics that existed at that time.

After many hours of chit chat, that night, my brother and I said our usual good by to our cousin, and after a good night sleep, we awoke bright and early on Sunday October 9, I983, went through our usual check list of things to do, and after securing the house, we went to our favorite local restaurant for breakfast. After breakfast we gassed up at a local gas station, and we were on our way back home to Connecticut.

During the return trip, we talked about our trip, and we also talked about some of the subjects we discussed during our rap session. Then out of the clear blue sky, like a bolt of lightening, the idea of a Two State Universe flashed into my mind. I immediately knew that this idea had merit, and it was correct, because I began to see answers to many of the conflicts and paradoxes that were known at that time. I told my brother about what had flashed in my mind. He agreed with me that the new idea could be used to answer all the questions we talked about the night before that seemed to have no solutions, plus some that had been bothering him.

We talked about the new idea all the way home. We were so deeply involved in discussing this new theory, that we passed our exit, on Route 84, by four stops, and we had to turn around to get back to our intended exit. When I returned home, I made a note about the new ideas that popped into my head and the date that it happened.

I decided that the ideas that flashed into my head were important and they should be worked on and put into a book. I spent many months working out the details of the new theory and solving many of the unsolved problems and paradoxes of science using the laws and conclusions that resulted from the new theory. I then began to write the book, which I titled, The Two State Universe. As I wrote the book, ideas continually flashed into my mind, and as I wrote the book, I felt as if I was reading the book instead of writing it.

I tried to write the book for the general reader; however, this could not be done. The book required that I coin new words, and it also required a considerable amount of mathematics to solve the problems described in the book.

The intent, therefore, of this book is to describe the new theory in a simple and interesting way without using any mathematics, so that anyone with an interest in science who can read, will be able to follow and understand The Two State Universe Theory that flashed in my mind. I am also adding new material that was derived after the book was written, and any new ideas that pop into my head, as I write this book.

# 2.
# The Two State Universe

The two-state universe theory is simple and beautiful. It proposes that the universe is neither a single state nor a three or four dimensional universe, but it is a two state universe in which everything that exists in the universe exists as a complement, and complements can exist only if they are kept separated in what man perceives as time and space.

Everything that exists in the universe, which is complex in design, appears to be made up of some type of building blocks. For example, our bodies are made up of billions of cells, atoms are made of charged particles, a television screen is made of many dots; and a color image is made of many red, blue, and green dots. Because of these observations, we conclude that the universe is also made of building blocks, consisting of an infinite number of points in space. Each of these points contain something or nothing, or the existence or nonexistence of something.

The universe is dynamic, which means that it is constantly changing, and everything that exists in it is constantly moving. The universe contains things that exist and things that do not exist, and because the universe is dynamic and everything in it is constantly changing, we conclude that each point in space at any one time could contain something or nothing or be in a state of existence or a state of nonexistence. These two conditions of existence or nonexistence are complementary conditions.

What do we mean by the words complements and two state? Complements can be understood to mean parts of the whole, and there are two parts to the complementary whole. The two parts are considered complementary if the truth of one means the other is false. For example, if it were true that at one point in space something exists, it would be false to state that at the same point at the same time nothing exists. And if it were true that at one point in space there is the nonexistence of something, then the existence of something at the same point at the same time would be false. Therefore, each point in space can be in one or the other complementary conditions, but they cannot be in both complementary conditions at the same time. Thus, each part of the complement is a state, and since there are two parts to the complement, and since a point in space can exist in either of these two complementary conditions, then each point must be capable of being in one or the other states. Because of this, each point in space must be two state in function, and since the universe is made up of infinite numbers of these two state points, then the universe operates as a two state universe.

Everything that exists in the universe operates on this two state principle, which means that everything has a complement in the two state universe. Examples of complements are an on and off light, an open and closed door, an open and closed meeting, an open and closed switch, winter and summer, hot and cold, man and woman, good and evil, a positive particle and a negative particle, a north pole and a south pole, something and nothing, real and unreal, true and false, happy and sad, wet and dry, a bounded and an unbounded universe, analog and digital, order and chaos, up and down, etc. I could go on and on because everything that you can think of has a complement.

In the two state universe, there are time separated. complements, such as the on and off light or the open and closed door complements. The light can't be on and off at the same time at the same point in space, and the door can't be open and closed at the same time. The on light is separated from the off light by time. The light is on at one time and off at a different time. The door is open at one time and closed at a different time. These are just two examples of the many time separated complements.

There are also space separated complements, such as the positive and the negative particle complement. The positive and negative particles can't exist in the same space at the same time because they are complements, and since they both exist at the same time, they can only exist separated in space. Another example of a space separated complement is the north pole and the south pole of a magnet. The north pole of the magnet can't exist at the same point in space at the same time as the south pole of the magnet, because they are also complements and they both exist at the same time, so they, too, must exist separated in space.

Thus, there are time separated complements and space separated complements in the two state universe, so we see that complements can exist only if they are kept separated in either time or space.

There are many and all kinds of complements. Some complements are something, and some complements are nothing. There are even complements that are part something and part nothing. The something nothing complement is just such an example of this kind of complement, because something is real and is and nothing is not real and is not.

The something nothing complement is important because it includes all of the above-described complements and any other complement that you can think about. See Figure 1.

In Figure 1 we have constructed a block diagram of the something nothing complement and some of its representative complements. The two parts of the complement are shown at the center of the drawing.

Connected to the something nothing complement is the existence nonexistence complement. Existence is something and it is connected to the something part of the complement. Non existence is nothing, and it is connected to the nothing part of the complement. Space and no space is another complement shown connected to the something nothing complement. Space is a void, and it is nothing, therefore, it is connected to the nothing part of the complement. No space means that something is occupying the space, therefore, it is connected to the something part of the complement.

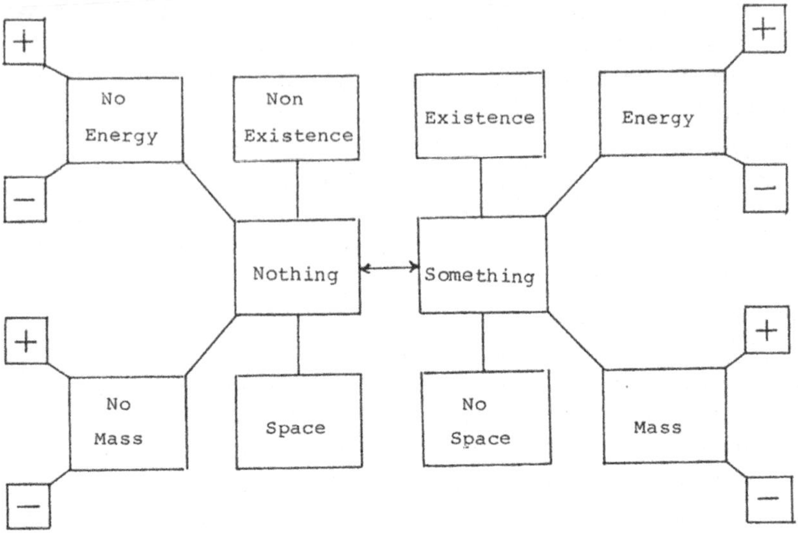


Figure 1.

Energy and mass are something, so they are connected to the something part of the complement. No energy and no mass are nothing and they are connected to the nothing part of the complement.

The figure also shows a plus and minus complement connected to the energy complement. This complement could represent the parts of a magnet. Every magnet has two well-defined regions that we call poles. One pole is called the north pole; the other is called the south pole. The magnetic field of the magnet contains energy. We define this type of energy to be pure or basic energy, because it differs from other types of energy, such as, kinetic energy and potential energy, which in the two state universe theory is considered to be complex energy.

Kinetic energy is the energy that a mass has as a result of its motion. However, it is made up of part mass and part energy, therefore, it is not pure or basic energy—it is complex energy. Potential energy is the energy that a mass has because of its position or separation; such as a book on the top of a table. When the book is on the top of the table it has potential energy because it has the potential of creating motion energy when it falls off the table. This energy is also not pure or basic energy because it is energy that is associated with mass. So It, too, is considered to be complex energy.

Magnetic energy, however, is pure energy, and the north pole and the south pole of the magnetic complement would be associated with energy; therefore, the magnetic complement is connected to the energy complement.

The mass complement also has a positive and a negative complement associated with it. This complement can be the positive proton and the negative electron complement. These complementary parts are the building blocks of all the atoms, and because individually they are basic building blocks of mass they are considered to be basic or pure mass as distinguished from complex mass. Each of the known atoms is complex mass because they are made up of pure mass, the positive and negative particle, and they also contain energy.

The mass of the earth is another example of complex mass because it doesn't consist of just pure mass or what is known as rest mass in physics, but it consists of pure mass and a considerable amount of energy. Therefore, the earth must also be classified as being a complex mass because it is made up of mass and energy. The earth is not a pure mass.

Associated with no energy is a plus and minus complement. This complement could be plus time and minus time because time is nothing; it has neither mass nor energy. Man invented it for his convenience to help him interpret the changes in states that are continually occurring in the two state universe.

There is also a plus and minus complement associated with the no mass complement. This plus and minus complement could be our number system, such as plus two and minus two or plus and minus any number. Numbers are also nothing because they have neither mass nor energy. They too were created by man for his convenience to help him interpret the real part of the universe.

The complements shown in Figure 1 are only a small example of the countless numbers of complements that can be associated with this complement, but they clearly show the numerous kinds of complements that exist in the two state universe.

The something nothing complement also shows that the nothing part of the universe is just as important as the something part of the universe. This complement shows that there is a something universe consisting of mass and energy and a nothing universe consisting of time and space. Both of these universes are complements existing side by side and interconnected in the two state universe. The two state universe theory requires that if we have a something universe, then we must also have a nothing universe, because everything that exists in the two state universe must have a complement, and the complement of a something universe is a nothing universe.

# 3.
# The G Factor

The digital computer and digital circuits, such as the **and**, the **nand**, the **or** and the **nor** circuits, could be used as a direct analog of complements because digital electronics is based on the complementary nature of the on off principle of a switch, a relay, a transistor, etc. In the digital circuit a 1 and 0 are complements in which the 1 represents the on state and the 0 represents the off state. This binary or two state system uses the transistor in these digital and computer circuits, because the transistor functions as a complement when it is operated in the switching mode.

When the transistor is operated in the switching mode it changes states from an on state to an off state. However, the transistor has a region known as the amplification region located between the off state and the on state of the transistor. Transistor amplifiers operate in this amplification region of the transistor. In this region the transistor is neither fully off nor fully on. In the two state universe, which operates on the complementary nature of points in space where the points can be in the something or nothing state or existence or non existence states, there must be a region or an instant in time when the point is neither in the fully something state nor in the fully nothing state. This region or instant in time is called the G point or G factor in the two state universe. This G point region of a point in space would be equivalent to the amplification region of the transistor. In the G factor complements can exist at the same point in space at a fixed interval in time. The G factor is an important concept in the two state universe, because creation is possible in the G factor.

In the two state universe creation can occur only in the form of complements, because everything that exists in the two state universe exists as a complement separated in either time or space. Therefore, the two complementary parts of the thing that is created must be created at the same interval in time. This condition can only occur in the G factor.

Does creation actually occur in the two state universe? There is a considerable amount of evidence showing that creation does occur in the two state universe. For example, in the something nothing complement we showed that there are plus and minus numbers associated with the nothing part of the complement, such as plus two and minus two. These two numbers are complements and they exist only if they are kept separated. If we combine them we find that plus two and minus two equals zero. By the reverse logic or by the principle of creation in the two state universe, zero equals the complement of plus two and minus two. Therefore, plus two and minus two was created from zero. This type of complement, however, is not real because neither of the two parts of this complement has mass or energy. They have no real existence—they are nothing.

This example shows evidence of creation; however, the complement created was from the nothing part of the complement. Is it possible to create complements that are real or from the something part of the complement? The two state universe theory suggests that it is possible to create something that is real and physical in the G factor, but the creation must be in the form of a complement.

There is strong evidence in physics which shows that real creation can and does occur in the labs of nuclear physicists because the nuclear physicists have already created real complementary particles in their labs, and as the two state universe theory predicts, the creation is in the form of complements. For example, in the lab a pair of electrons were created, one being a negative electron and the other was a positive electron called a positron. This creation occurred during the collision of an electron at rest by a high-energy fast moving electron, resulting in four electrons from two electrons. However, one of the four electrons was positive, which means that a complementary pair of electrons were created. These particles were short lived because they were annihilated within split seconds of their creation when the complementary pair of electrons, because they were attracted to each other, came together in the same point in space at the same time.

Also, at the University of California at Berkeley, a proton antiproton complement pair was created when a stationary proton contained in hydrogen was bombarded with a beam of protons. The resulting reaction was the creation of three positive protons and one negative proton, called an anti-proton. These two created real particles were also annihilated within a split second of their creation. This reaction was similar to the one described above for the creation of the electron pair.

These reactions are predictable from the two state universe theory, and as predictable, real particles were created in the lab. They were created in the form of complements at a point in space during a split second in time. These laboratory results are amazing and strong evidence for the existence of the G factor in the two state universe.

# 4.
# The Three and Four Dimensional Universe

The two most popular theories of the universe are the three dimensional universe theory and the four dimensional time-space universe theory. The three dimensional theory proposes that everything that exists, exists in a three dimensional space. In this theory space is considered to have three dimensions of length, width and height. The four dimensional time-space universe theory proposes that everything that exists, exists in space-time in which space has the three dimensions of length, width and height, and time is considered to be the fourth dimension of the universe.

The problem with both of these theories is that space is nothing and time is nothing. Neither space nor time has mass or energy—they are absolutely nothing. It is difficult to understand how nothing can have any properties, such as length, width and height or how time, which is nothing, could be anything but a concept of man's thought. Also, neither of these theories has any laws, ways or means to handle the puzzling paradoxes and riddles that occur in the three or four dimensional universe.

The two state universe theory is a completely different concept and operates in an entirely different manner. It also has the ways and means for handling or avoiding the puzzling paradoxes and riddles that occur in the three or four dimensional universe.

In the three and four dimensional universe theories, everything that exists is considered to exist as single or separate entities in the three dimensions of space or the four dimensions of space and time.

However, everything does not exist as single entities, they exist as complements, and as we have seen from the two state universe theory, these complements are interconnected and must abide by specific rules and laws to exist. If we look at everything that exists in the universe as separate entities without considering them to be complementary, our view of the universe will be false and distorted resulting in illogical and paradoxical problems that appear to have no solutions. For example, we have seen that we cannot have the complementary parts of a switch existing separately, otherwise we can have a switch that is on and off at the same time, and if we treat the complementary parts of a door separately, then we can have a door that is open and closed at the same time. In a space separated complement like a magnet, we cannot treat the complementary parts separately otherwise we could have a single pole magnet or we could have a magnet with two north poles or two south poles, or we can assume incorrectly that the two complements can exist in the same place at the same time.

These are examples of complements that are simple and obvious for man to comprehend and understand that the complementary parts cannot be treated as separate entities. However, there are many complements that are treated as separate entities in the three and four dimensional universe that are not easily understood to be complementary, such as the irresistible force and the immovable object, mass and energy, the chicken and the egg to name a few. We will examine these complements in detail later.

Thus we see that the three and four dimensional universe theories simply don't include or consider the complementary nature of the universe, and they don't contain the rules and laws to show how the complements exist and how they operate.

Because of this we see how easily we can get into trouble in the three or four dimensional universe by wrongfully concluding that the complementary parts of the complements exist as separate entities. Our wrongful conclusions about the complementary nature of the universe generally results in circular reasoning about the complements which creates paradoxes that appear to have no solutions. This is why there are so many paradoxes, riddles and unanswered questions puzzling the scientists today.

Therefore, we see that the two state universe theory makes more sense than the two popular theories of the universe, because it not only allows us to be aware of the complementary nature of everything, as they really exist, but it also gives us the laws for understanding how they operate. Also, it doesn't require that we give dimensions and properties to nothing, and it doesn't allow us to conclude that time, which is also nothing, can vary with velocity. It is difficult to understand how nothing can vary with velocity.

In the two state universe theory time is nothing but our way of perceiving the changes in states that are constantly occurring in the universe, and measurements are made among and between the things that exist.

Someone might ask, how can space or time, which is nothing, separate the complements. Well, it is not nothing that keeps the complements separated; rather it is the forces that are acting on the complements that separates them. It only appears that nothing (space and time) is keeping the complements separated.

Two types of forces act on the complements. One of the forces acting on the complements is the force that keeps the complements separated in order that they can exist. These forces are the forces of existence or existence forces. Centrifugal force would be an example of an existence force. For example, it is the centrifugal force of the revolving electron in an atom that keeps the electron separated from the positive proton or nucleus, which allows the complement to exist. There are also forces of attraction between the complements that works to bring the two complementary parts together. These forces are attraction forces or annihilation forces that can cause the complements to come together in the same point in space at the same time resulting in their annihilation. The electrostatic force of attraction between a positive particle and a negative particle would be an example of an attraction force. All real complements are attracted to each other. Therefore, it is the balance of the forces of existence and the forces of attraction between the complements that keeps them separated, not the nothing of space and time. It only appears to man that time and space is keeping the complements separated.

In the following chapters we will solve many of the paradoxes, riddles and problems that exist in the universe today because of the three and four dimensional viewpoint of the universe. We will solve these problems by using the laws, postulates and conclusions developed from the two state universe theory. These laws, postulates and conclusions are listed in the appendix.

# 5.
# Infinity

Infinity means unlimited or unbounded. One of the great mysteries of our universe is the concept of infinity, which implies the concept of an unbounded universe. It is difficult to imagine a universe that goes on forever without any bounds, but it is just as difficult to imagine a universe that is bounded, because if it is bounded what is it bounded by or what bounds it?

These two conditions of bounded and unbounded are complementary conditions. This paradox results from the three dimensional or even the four dimensional time-space universe theory viewpoints, because both of these theories considers both of these complements, the bounded and the unbounded universe to be separate entities. However, in the two state universe we recognize that both of these two conditions are not separate entities, rather they are complementary conditions, and they both must exist in our universe. That is, if the universe has an unbounded state it must also have a bounded state, because everything has a complement in the two state universe. How can this be? Let us look at the problem closer and see if we can make sense out of this conclusion, which is required from the two state universe theory.

Let us try to solve this problem by using what some scientists call a thought experiment. In this experiment we will assume that our most brilliant scientists invented two robot brothers capable of living forever. One of the brothers we will call up and the other we will call down. If the up brother takes an imaginary rocket trip straight up in the universe and the down brother takes an imaginary rocket trip straight down in the universe, to determine once and for all if the universe extends forever or if it has some kind of bounds; then, as the two brothers travel on their journey they will both pass complementary points in their travel, such as plus two minus two, plus fifty minus fifty, or any number, such as plus N minus N. The up brother reaching the plus complementary distances and the down brother reaching the negative complementary distances.

These distances could represent distances of meters, miles, light years or any other measurement of distance. Eventually they will both reach a distance in the universe where there is no more mass and energy, and they will then enter the nothing part of the universe, space and time. When they both reach the nothing part of the universe, they could both travel along the edge of the something universe where they could meet and pass each other.

In this case the up brother could return home by the path that the down brother traveled. When he returns home he will still be traveling in the up direction. Likewise the down brother could follow the up brother's path and return home still traveling in the down direction. Therefore, when the brothers return home they will be traveling in the same direction they were traveling in when they started their journey.

In this case the brothers continued to travel in the bounded universe. They could have both continued to travel in the direction they were headed in when they arrived at the nothing universe. In that case they would have entered the nothing universe. However, when they enter the nothing universe they would in effect be extending the bounds of the something universe, and since the nothing universe doesn't have any dimensions or characteristics, the brothers could continue to travel in the nothing universe, extending the bounds of the something universe as they travel on to infinity. The up brother going on forever to plus infinity, and the down brother going on forever to minus infinity.

Another way of looking at this same problem of the bounded or the unbounded or infinite universe is to realize that in the two state universe we have two complementary universes existing side by side—the something universe, made up of mass and energy, and the nothing universe, made up of time and space.

This means that if the two brothers take a rocket trip, one going up and the other going down, they will eventually reach the end of the something universe. When they reach the end of the something universe they will have reached the end of the bounded universe. The end of the something universe is the bounds of the real universe because there is absolutely nothing beyond it. However, if the brothers want to continue beyond the bounds of the something universe, they can because when they enter the nothing part of the universe, the very fact that they are something, means that they can extend the bounds of the something universe for as long as they wish, simply by continuing to travel in the unbounded or nothing part of the universe.

If they stop their rockets at any point during their journey in the nothing universe, there will be nothing beyond their rockets, and since nothing is nonexistence there would be non-existence beyond the point where they stopped. Therefore, we conclude that the universe is bounded and unbounded, as predictable from the two state universe theory.

Many scientists have difficulty understanding the concept of infinity; they don't know what it is. In the two state universe we see that the concept of infinity exists and results because of the two state nature of the universe. Infinity exists in our universe because the universe is complementary; there is a something universe which is bounded and a nothing universe that is unbounded. Infinity exists because of and because it is part of the unbounded universe. There is no end to an unbounded universe because it consists of absolutely nothing, and nothing can't have dimensions or any characteristics. Infinity is merely a symbolized way to describe this endless and dimensionless unbounded universe. It describes the endless space when all the points of space are in the nothing mode. It exists as a space separated complement of plus infinity and minus infinity as part of the nothing complement.

Another term that results because of the nothing or unbounded universe is eternity. The nothing universe is made up of the complement of space and time. Time is nothing, and it is also unbounded. It can, therefore, go on forever. It is eternal. Thus the concept of eternity. Infinity results from and is associated with the unbounded space, and eternity results from and is associated with the unbounded time.

# 6.
# Zero

The concept of zero, as we will shortly see, is also mysterious and paradoxical. Zero can be considered to be the complement of plus infinity and also the complement of minus infinity because zero is minus with respect to plus infinity and it is plus with respect to minus infinity. Therefore, although plus infinity and minus infinity are complements, they can be considered to be a pair complement consisting of the complements of minus infinity and zero and zero and plus infinity, and since zero contains both the plus infinity complement and the minus infinity complement within its bounds, it can be considered to be a G factor point and also a point in space, which, as we have seen, is two state in function.

Zero means absolutely nothing or having no magnitude, or it can be the point from which all kinds of measurements are made. It is represented by the symbol 0. Let us examine some of the paradoxes and problems we encounter when using zero as the reference point of a measurement.

If we have a yardstick of thirty-six inches and we cut the yardstick in half, we will have eighteen inches. If we cut the half yardstick in half again we will have nine inches, and if we cut the nine inches in half, we will have four and a half inches. Although there is a half an inch involved, it could still be cut in half, and we could continue to cut the fractional parts of the yardstick in half.

A paradox results if we ask the question Will we ever reach zero by continuing to cut the stick in half? The process of continuing to cut the stick in half results in a paradox because we will always have a half left to cut. That is, no matter how many times the stick is cut, there will always be a half of the last cut remaining. Logic tells us that we should be able to reach zero by cutting the stick, yet we can't reach zero because we will always have something left to cut. What is the answer to this paradox? The answer can be found by means of the two state universe theory; because if we continue to cut the stick we will eventually reach a point where we can't make any further half cuts. At this stage we will have reached what the two state universe theory calls a point in space.

A point in space is two state in function, because it can be either something or nothing. In the two state universe a point in space can't be cut in half. No matter how hard we try, we can only cut the point at the beginning or the end of the point. Therefore, we reach a point in space that is separated from zero by what mathematicians call a differential distance. This seems to indicate that we don't reach zero. However, in the two state universe, the complement of "not reaching zero" is "reaching zero," and both of these complements must exist. Therefore, how do we reach zero? We can accomplish this by means of the G factor. If we try to cut the point in space in half we enter the G factor where zero and the differential distance both exist during an instant in time, but the two state universe is constantly changing; therefore, the point can go to zero or back to the differential distance in the G factor. If it goes to zero we will reach zero by the division. If the point changes states and ends up the differential distance we do not reach zero. Therefore, we see that the complement of reaching zero and not reaching zero does exist, as it should in the two state universe. Although we can reach zero by division, there is still a problem with the concept of cutting a distance in half continuously to reach zero. Let us examine this problem in a slightly different way.

Let us assume that instead of the yardstick of thirty-six inches, we have thirty-six points in space. If we cut the thirty-six points in half we will have eighteen points, and if we cut the eighteen points in half we will have nine points. However, if we now try to cut the nine points in half, we will reach the four and a half point, but we can't cut a point in half; therefore, we will either reach the fifth point or the fourth point depending upon the state of the point. If we end up on the fourth point we can then continue to cut the four points in half to reach two points and we can cut the two points in half and we will again reach a point in space, which we have already discussed above; and as we have seen, we can reach zero by division if the point in space is in the zero or nothing state, because zero and the differential distance are complements of the point in space.

If, however, instead of reaching the fourth point we reach the fifth point, our next cut would give us either two points or three points, and the same logic described above would continue. We can see from this analysis that in this case the cutting was discontinuous or what the scientists call quantum in nature. This means that the universe is grainy in nature, when we reach the very small or a point in space.

When we deal with larger measurements, the grains, although they exist, are not apparent, and in this part of the universe things appear to be continuous, such as when we were cutting the yardstick in half. The yardstick could apparently be cut in halves until we reach the very smallest cuts as we approach the point in space. When we reach the point in space the cut can easily be seen to be discontinuous. Therefore, we see the continuous and the discontinuous complement or the classical and the quantum nature of the universe.

This problem is interesting because it not only showed how the two state universe theory handled these paradoxical problems, but it also demonstrates that if we developed a science of classical physics where forces, distances and energy are considered continuous, we would realize that there must also be a science where things are not continuous (i.e. where things are discontinuous). This field of science where things are discontinuous is called quantum mechanics.

Also, today there is a theory of physics called the Chaos Theory. This theory could have been predicted from the two state universe theory, because if there is a science of order, then there must be a science of disorder or chaos.

Thus we see that the concept of zero and infinity are important concepts in the two state universe theory, and we also see how much more simple these concepts are when we view them from the two state universe theory, than when we view them from the three and four dimensional universe theories, where these concepts are difficult if not impossible to comprehend.

# 7.
# Two Riddles

1. The Chicken or The Egg Riddle

The riddle of what came first the chicken or the egg is an interesting riddle that has been around for a very long time. The riddle exists because neither the three dimensional nor the four dimensional time-space universe theories recognize the complementary nature of the universe, and neither of these theories have any laws, rules or postulates for solving this problem. This riddle will continue to exist if we continue to view it from the three dimensional or the four dimensional time-space universe theories.

The two state universe theory has postulates, rules and laws for solving this problem, and in fact it's a rather simple problem to solve from the two state universe theory.

Let us examine the riddle first from the three and four dimensional theories. If we examine it from the three or four dimensional universe theories, our reasoning will be something as follows: If the egg came first, where did the egg come from without a chicken? If the chicken came first, where did the chicken come from without an egg? This kind of reasoning is circular reasoning, which leads to a paradox, because neither answer seems to give a solution. If we examine the riddle closer, we may reason that maybe by some miracle the chicken came first. Would this reasoning answer the question? The answer of course is no because even if the chicken came first the chicken would lay an egg but the egg could not be hatched, because it would not be fertilized. If the egg came first it would not hatch if it was not fertilized, but if it was fertilized it would hatch and we would get either a rooster or a hen. If it hatched into a rooster the cycle would end because the rooster couldn't lay an egg.

If the egg hatched into a hen we would be back to where we started from, because if the hen laid an egg it wouldn't be fertilized; therefore, the cycle would end. This circular reasoning results in the riddle of the chicken or the egg, which appears to have no solution.

However, looking at the problem from the two state universe theory, we have two postulates, derived from the new theory, that we can use to solve the riddle. What are these two postulates?

Postulate 5. Creation can occur in the form of complements.

Postulate 6. Creation is possible in the G factor. See Postulates 5 and 6 in the appendix.

Using both of these postulates, the solution can be readily determined, because we immediately realize that there must have been the complement of the rooster and the hen created at the same time in the G factor at the time of their creation. With a rooster and a hen, the hen can lay eggs that will be fertilized and they will hatch into roosters and hens to continue the cycle. Someone might ask why we assumed that the rooster and the hen were created in the G factor instead of two eggs. Well, from the postulates we can see that neither one nor two eggs can be created in the G factor because neither of these conditions is complementary. Only two complementary fertilized eggs can be created in the G factor. This means that one of the eggs would have to be fertilized a rooster, and the other egg would have to be fertilized a hen; and these two complementary fertilized eggs are the full equivalent of a rooster and a hen, because the fertilized eggs are merely different degrees in the life of the rooster and the hen. Therefore, the obvious solution to the riddle of the chicken or the egg from the two state universe theory is that the rooster and the hen came first.

## 2. The Irresistible Force and the Immovable Object Riddle

The problem of what happens when an irresistible force meets an immovable object is another riddle that causes circular reasoning from the three or four dimensional time-space universe theory. The circular reasoning occurs something like the following: If we reason that an irresistible force is a force that nothing could stop; then, we would conclude that nothing could stop this force, not even the immovable object. However, by the same reasoning if we assume that nothing could move the immovable object, not even the irresistible force, we would realize that this reasoning presents two logical answers neither of which gives the solution.

The more we ponder this question the more confusing it gets. We could come to all kinds of conclusions, such as the irresistible force will win out, or we could conclude that the immovable object will not move, and it will win out. Maybe they will destroy each other. This problem is paradoxical, and it doesn't have a solution from the two popular theories.

However, this is another simple problem viewed from the two state universe theory, because we immediately realize that these two conditions, the irresistible force and the immovable object, are complementary. This means that these two complementary parts can never meet or come together in the same point in space. They must be kept separated in either time or space in order to exist.

This problem would be equivalent to asking the question, what happens when a light is on and off at the same time? Or what happens if I go up and down at the same time? This problem arises because we think of the irresistible force existing as a separate entity from the immovable object, and as we now know, this is not the way the universe operates. These two riddles and their solutions clearly demonstrate the complementary nature of our universe, and they strongly support the two state universe theory.

# 8.
# The Dual Nature of Light

The dual nature of light controversy, which still exists today, is another paradox. It began with two logical and scientifically sound theories: Newton's corpuscular theory and Huygen's wave theory of light. Since both of these theories are logical and appear to be scientifically sound, scientists are at a loss to explain light. Is it a particle or is it a wave? Further, why should there even be two theories of light? We not only have two theories of light, but as we will soon see, these two theories are in conflict with each other.

Newton conducted experiments showing that if sunlight is passed through a triangular prison the light would separate into the different colors of the rainbow, and because he was able to separate white light into the different colors of the rainbow, he reasoned that, if he could pass these different rainbow colors through a second prism, they should recombine and form the original white color of light. Using a second prism he passed the rainbow of different colored lights through this second prism, and as he expected, they did indeed recombine into the original white color. From this experiment he reasoned that light must be made up of different colored particles called corpuscles.

In 1678 Christian Huygens, physicists, mathematician and astronomer, proposed a different theory of light. He believed that light is wavelike, not a stream of colored particles. He believed that light traveled through space in the form of wavefronts in which each point on the wavefront was a point source of light. Each of these point sources of light were like little spotlights that would radiate light in a forward direction creating new wavefronts as the light traveled through space.

Newton and Huygens believed that light traveled in a straight line. However, Francesco Maria Grimaldi (1618-1663) experimentally showed that this was not true. He found that light spreads out after passing through a hole. This spreading out of the light he called diffraction. Newton and Huygens were both aware of Grimaldi's experiment, but neither one paid any attention to it. Over a hundred years later, Jean F. Fresnel (1788-1827), using the wave theory of Huygens, was able to show that diffraction of light was due to the wave nature of light.

Up until this time, both theories were considered to be scientifically sound and equally possible, but because of the conflicts between the two theories, the scientists were unable to determine which of the two theories were correct. Newton's theory could explain how the multicolored corpuscles could travel through the vacuum of space, but it couldn't explain the spreading of light; Huygen's theory could explain the spreading of light, but it couldn't explain how the energy of light in the form of a wave could travel through the vacuum of space.

It is well known that energy in the form of a wave needs a medium to travel in, because the energy must wave something in order to travel in space. If I talk to someone, my vocal cords vibrate the air in the room. This vibration travels to the listener’s ear where the vibrating air causes the listener's eardrum to vibrate. Likewise, if I drop a pebble in a pond, the energy of the dropped pebble will generate a water wave that travels out in all directions from the point where the pebble was dropped in the water until the wave reaches the shore where the energy is expended. If an object is floating on the surface of the pond, the energy of the water wave will cause the object to vibrate when the wave passes the object.

However, a light wave traveling from the sun to the earth must pass through the vacuum of space, but in the vacuum of space there is nothing to wave. How then can the energy of light reach the earth through the vacuum of space where there is no medium for the light wave to vibrate? How can we explain this conflict? Well, the scientists at that time postulated that there must be an aether space which was something for the wave to vibrate but nothing for material bodies to travel in, otherwise, if it was something for material bodies, it would cause resistance to the moving bodies and slow them down.

Although they could not explain this strange aether, they reasoned that it must exist otherwise the wave of Huygens could not travel through the vacuum of space.

The aether was a very strange substance because it had to be something, but it also had to be nothing. This is another paradox created from a three or four dimensional universe viewpoint. However, from the two state universe theory, this strange aether sounds like a vacuum of space consisting of infinite numbers of two state points that are nothing for the material objects but something for the light waves to travel in.

The dual nature of light controversy continued until Thomas Young (1773-1829) discovered the principle of interference. Interference occurs when two waves meet at a point. If the two waves are light waves and they meet at a point on a white screen vibrating in the same direction, then the two waves add together and the point where they meet will become light, but if the two light waves meet at the screen vibrating in opposite directions, then the two waves will subtract, causing the point where they meet to become dark. If the waves add, this is called constructive interference; if the waves subtract, it is called destructive interference.

Young made two narrow slits in an opaque screen. Then he allowed sunlight to pass through a pinhole onto the opaque screen. After passing through the two slits, he allowed the light to fall on a light sheet of paper set some distance from the opaque screen. The result was an interference pattern of light and dark bars on the light sheet of paper, caused by the constructive or destructive interference of the light waves where they met on the light sheet of paper. This experiment, with the effect of diffraction proposed by Grimaldi and Fresnel, clearly indicated that light was a wave.

Then in 1864, James Clerk Maxwell (1831-1897) published a paper that mathematically analyzed the work and laws of Michael Faraday (1791-1867). The equations presented in the paper were known as Maxwell's equations. These equations showed that an electromagnetic wave traveled through space at a speed equal to $3 \times 10^8$ meters per second. This is the exact speed that light travels through the aether. He concluded that the two speeds were too identical to be a coincident; therefore, he reasoned that light must be an electromagnetic wave. This conclusion means that light is not mechanical but electromagnetic in nature. He also concluded that an accelerating electron generates a magnetic field, which in turn generates an electric or electrostatic field, etc. The two fields travel at right angles to each other, and the wave travels perpendicular to both. From this finding he further concluded that this must be the way that light travels through the luminiferous (light bearing) aether. This was the name Maxwell gave to the aether.

His work further shows that light is a very narrow band of electromagnetic waves with each color consisting of a particular frequency within the total bandwidth of frequencies making up the spectrum of visible light. Also, since light is an electromagnetic wave and an electromagnetic wave is a transverse wave, it was concluded that light must also be a transverse wave. This conclusion was verified by experiments with polarized light. A transverse wave is a wave that vibrates at 90 degrees to the direction of travel of the wave. It differs from a longitudinal wave, which is a wave that vibrates back and fourth along the direction of travel. A sound wave would be an example of a longitudinal wave.

About 25 years later, Rudolph Hertz (1857-1894) showed that indeed an accelerating charged particle radiates energy, as described by Maxwell, in the form of an electromagnetic wave. In 1888, using Maxwell's findings, he transmitted an electromagnetic wave through space. Ten years later, Guglielmo Marconi (1874-1937) developed the radio, and in 1901 he sent a message across the Atlantic.

Although the wave theory of light had become well established, some scientists were still not comfortable or satisfied with the idea of an aether space, which was merely a hypothetical medium created by the scientists to help them explain how energy in the form of a light wave could travel through space. Since the aether appeared to be nothing yet it had properties, it was impossible for the scientists to comprehend it, until Michelson and Morley came along and conducted their famous experiment.

Albert A. Michelson (1852-1931) first by himself and later with the help of Edward Morley (1838-1923) designed an experiment which they believed could detect the elusive aether. Knowing that the earth traveled around the sun at 30 kilometers per second, they believed that an interferometer would be capable of detecting the aether. However, their experiment failed, which led them to conclude that there is no aether. This conclusion further complicated the problem of the dual nature of light. We will discuss the Michelson Morley experiment in more detail later.

# 9.
# The Controversy Continues

During the late eighteen hundredths and early nineteen hundredths, not only did the dual nature of light controversy continue, but dramatic new ideas in science were being developed. This period in the history of science is considered to be the beginning of Modern Physics, which produced dramatic new changes in physics, which up until that time was known as classical physics.

During this period, Albert Einstein (1879-1955) developed his famous theory of relativity, which consisted of two parts: the special theory of relativity in 1905, and the general theory of relativity in 1915. Both parts of the relativity theory were in conflict with Newton's theories. The special theory of relativity was in conflict with Newton's laws of motion, particularly Newton's laws of motion as they applied to a mass in motion approaching the speed of light. Newton's laws assumed that mass was a constant, and Einstein's theory, among other things, proposed that mass increased with velocity. Einstein's general theory of relativity was in conflict with Newton's law of gravity. Newton assumed that masses attracted each other, because of an action at a distance. However, Einstein assumed that space-time was curved or warped by the presence of a mass, and two masses came together because they followed the path of least resistance in the curved space-time.

Although some scientists claim that Einstein was not aware of or influenced by the Michelson Morley experiment when he proposed his theory of relativity, a 1923 English translation of his original paper on the principle of relativity, published by Dover Publications, Inc., seems to indicate that he was aware of their experiment. Because in the second paragraph of his paper, Einstein makes reference to, " . . . the unsuccessful attempt to discover any motion of the earth relatively to the "light medium," suggest . . ." It is clear from this statement that Einstein was aware of the Michelson Morley experiment when he wrote his paper on the special theory of relativity in 1905.

Also, around this same time, another revolutionary new theory was beginning to be developed that would further complicate physics. However, the new theory, which would become known as quantum mechanics, was not developed in a short time by the brilliance of one scientist; but it was developed over a period of time by the work of many scientists. Many of whom were close friends who helped each other with their ideas.

The idea of the quantum theory began about 1900 when Max Plank (1858-1947) proposed that radiant energy could not be radiated or absorbed in the form of continuous waves; rather it must be radiated or absorbed in units called quanta (or photons). The energy of the quanta being proportional to the frequency. The idea that only a particular amount of energy could be radiated or absorbed suggested a discontinuous nature of energy; rather than the continuous nature of energy viewed by classical physics.

Plank discovered the quantum of energy while working on the problem of thermal radiation. A frequency spectrum of thermal radiation showed that although there was some energy in the high frequency violet region of the spectrum as well as the low frequency region there was an unbalance in the amount of energy contained in these different regions of the spectrum. To explain this problem, Plank discovered that if he quantized the energy, by assuming that energy was radiated in the form of fixed bundles of energy, he could explain the unexpected energy spectrum of thermal radiation. He was able to get good agreement between his empirical formula and the actual experimental curve of thermal radiation by assigning a value of $6.63 \times 10^{-34}$ joule-second to the proportionality constant, called h, in his quantum energy equation showing energy proportional to frequency.

During the same period it was discovered that if a negative charged zinc plate was illuminated, it would give off electrons. It was also discovered that the electrons were given off more easily with ultraviolet light, but not so readily with red light. Further, a brighter light was found to give off more electrons. However, for a particular frequency of light, all the electrons given off had approximately the same velocity. This effect is contrary to what we would expect from classical physics, because the brighter the light the more energy it would have. And, with more energy applied to the electrons in the metal, the electrons should escape from the metal with a higher velocity. But brightness has little effect on the velocity of the electron.

Einstein, realizing that this effect must be due to the quantum nature of energy, proposed a solution to this unexpected result by stating that the electron always absorbed only a quantum of energy. And if this energy is sufficient to cause the electron in the metal to overcome the surface barrier forces of the metal, then the freed electron would have a kinetic energy equal to the quantum of energy given to the electron minus the energy lost in getting the electron to free itself from the surface barrier of the metal. Since the electron only acquires a quantum of energy, it could only acquire a particular value of velocity. In 1922, Einstein received the Nobel Prize, not for his theory of relativity, but for his work on the photoelectric effect.

With the discovery of the photon, the corpuscular theory of light was revived. Because light traveling through space in small packages of wavelike energy is corpuscular in form. A. H. Compton (1892-1962) demonstrated this corpuscular form of a photon experimentally in 1921. The experiment showed that when a photon collides with an electron, the photon behaves like a particle. He was awarded the Nobel Prize in 1927 for his findings.

There were many more exciting discoveries being made around the turn of the century, which further complicated the dual nature of light controversy. In 1897, Joseph John Thomson (1856-1940) also referred to as J. J. Thomson experimentally discovered the electron. In describing the electron he referred to it as a corpuscle, which at that time was the way a small particle was defined. As a result of this discovery, he reasoned that this small negative particle must be one of the building blocks of the atom.

Thomson knew that matter was made up of atoms and atoms don't have a charge, they are neutral, so he reasoned that the atom must also contain a positive particle to cancel out the negative charge of the newfound electron. Therefore, he proposed a theory that the atom was composed of a sphere of positive charge in which the electrons were imbedded. He described this model of the atom to be like plum pudding.

Earnest Rutheford (1871-1937), a protege of Thomson, agreed that the atom may be constructed this way; but through work done by his assistant Hans Geiger and an undergraduate named Earnest Marsden, Rutherford discovered that the plum pudding theory of Thomson may not be correct. Because on bombarding a gold leaf with alpha particles, he noticed a significant scattering of the alpha particles, which could only be accounted for if the positive particle (the proton) was concentrated at the center or nucleus of the atom. That is, a significant number of the alpha particles were being deflected at such sharp angles that these angles could not be accounted for on the basis of the plum pudding theory. In fact, some of the alpha particles were being bounced backwards toward the direction they came from. From this work Rutherford concluded that the atom must contain a nucleus of positive particles. He calculated the size of the nucleus to be less than $10^{-15}$ meters in diameter.

Later, Neils Bohr (1885-1962), who also worked with J. J. Thomson, proposed a new theory of the atom that was based on Rutherford's nuclear atom and the quantum nature of energy. Bohr realized that, using the principles of classical physics, an electron would be able to revolve around the positive nucleus in any orbit, but because of Maxwell's finding that a revolving electron radiates energy, he reasoned that the electron would rapidly lose energy and spiral into the positive nucleus. He further reasoned that the quantum nature of energy must somehow account for the fact that this does not occur in the atom.

With the advent of spectroscopy, all kinds of spectrum lines of the elements were obtained and studied. One was obtained for the hydrogen atom called the Balmer series. When Bohr learned about the Balmer series from a friend, he immediately realized that this information was the missing piece needed to complete his theory of the atom.

Bohr knew that if the orbits of the revolving electrons were restricted a line spectrum, such as the Balmer series, would result. He also believed that the electron would not radiate energy in each of the allowed orbits. The latter being circles in which the radius of the orbits increased as the square of the whole numbers 1, 2, 3, 4, etc. For example, the first orbit would be 1 multiplied by itself, which is 1; the second orbit would be 2 multiplied by itself, which is equal to 4 etc. And the energy of each orbit would be inversely proportional to the radius in a negative manner, meaning that the energy increases with a larger radius (i.e. the higher orbits).

When the atom receives energy from light or any other source, the atom absorbs the energy and raises the electron to a higher orbit. The electrons are in an excited state in the higher orbits. On returning to the stable or lower energy state, the electron emits the energy it absorbed, and if the wavelength of this energy were within the visible range light would be given off.

The model of Bohr's atom successfully accounted for the Balmer series formula and the constants of the formula, which were accurately determined by spectroscopy.

The Bohr atom, therefore, showed that when a quantum of energy is given up we get a line spectrum of light. The quantum being the energy needed to raise the electron from one orbit to another. This quantum nature of the atom was also helpful to show the quantum nature of light, which further helped to revive Newton's corpuscular theory.

Although Bohr's atom successfully accounted for the Balmer series, Bohr was unable to explain why the electrons rotated around the nucleus in particular orbits.

In 1924 Louis de Broglie proposed a solution to this problem, which later won him the Nobel Prize. De Broglie reasoned that if energy in the form of a wave is particle like, then a particle of matter, such as protons, neutrons, and electrons, must be wavelike as well as corpuscular. He was able to show this by deriving a formula for the wavelength of these particles by equating Einstein's equation for the energy of mass with the energy of the photon.

Any wave that repeats itself would have a wavelength that is equal to the measurement from the beginning of the wave to the end of the wave. The frequency of the repeating wave would be the number of waves occurring in a particular time, such as the number of waves occurring in one second. For example, if a wave repeated itself 60 times in one second, then the frequency of this repeating wave would be 60 cycles per second or 60 Hertz, and the wavelength of this wave would be the distance between the beginning of the wave and the ending of the wave.

De Broglie proposed that an electron being wavelike would rotate around the nucleus at any radius or orbit that could hold a whole number of wavelengths. That is, only whole numbers of wavelengths must fit evenly within the circumference of the orbit. Therefore, during each successive revolution the electron wave would always be in phase, meaning that its path would be exactly the same during the second and succeeding revolutions as it was during its first revolution. In this manner the wave reinforces itself.

De Broglie's theory was expanded on by other scientists, such as Schrodinger and Heisenberg, whose mathematical theories on the wave nature of the particle became known as the wave mechanical theory or wave mechanics. Schrodinger and Heisenberg both won a Nobel Prize for their work on quantum mechanics.

Erwin Schrodinger (1887-1961) in 1926 derived a single partial differential equation that describes the wave equation of the electron in orbit, the de Broglie wave and the de Broglie rule of momentum which relates the momentum of an electron to the number of waves in a meter of the wave that describes the electron in orbit. Schrodinger visualized the electron in the atom as being made up of a number of waves interacting with each other. His equation, therefore, describes the electron orbit as being a sort of cloud. Although these equations shed some light on the atom and on Planck's constant, they fail to explain how a particle described as a wave can be contained within the space of an atom, because as we know, a wave naturally spreads out in all directions and is not bounded. How can a wave or combinations of waves be contained within the small space of an atom?

Schrodinger's equation was received favorably by the physicists because they could easily understand a quantum leap from orbit to orbit as being combinations of waves going from one combination mode to another in a continuous manner; rather than visualizing an electron jumping from one orbit to another in a discontinuous manner.

The discovery that mass could be both a particle and a wave led Werner Heisenberg (1901-1976) to his theory of the uncertainty principle. Heisenberg believed that it would be better to try to understand the quantum nature of the atom not by trying to visualize electrons jumping from orbit to orbit but by working with the spectrum of the atom, because the spectrum could be accurately determined from a spectrograph, and the spectrum could be seen.

He originally wrote equations based on the frequencies obtained from the spectrograph. And because there were a number of these equations needed to describe the spectrum of an atom, they were referred to as the quantum mechanical series. However, Heisenberg had difficulty interpreting these equations until he realized that the observer plays an important role in the measurement of small particles, such as the electron. And after reducing these equations to a single equation, his equation mathematically describes the uncertainty principle, which, in effect, states that it is impossible to exactly determine both the position and the momentum of an object or anything connected to its momentum. On applying the uncertainty principle to measurements of small masses, we encounter the problem that our measuring devices affect either the position or the momentum of the mass we are trying to observe. When the mass is great the momentum would be great and the error insignificant, but when the mass is small the momentum is small, and it leads to an error as suggested in the formula he derived from the quantum mechanical series. If the position is precisely determined, then the error in determining the momentum would be infinite, which means that we can't measure precisely both the position and the momentum of a small particle at the same time.

According to Heisenberg we are prevented from examining any event that is so small in size that the fact we are attempting to measure it we disturb its momentum or position. We arrive at types of measurements we are forbidden to make; otherwise our errors become so great that our observations become false. As a result, we are prevented from observing the wave nature of light at the same time that we observe the particle nature of light.

Led around in a circle, it appears we have reached a dead end. If we can't measure and observe the dual nature of light, will we ever be able to unlock its mystery? Will we ever be able to answer the questions: What is light? Is it a wave or a particle? Was Newton right? Was Huygens right? Is there an answer to this paradox?

# 10.
# The Wave Particle Paradox

The dual nature of light controversy is a good example of how a paradox results when two complementary parts are viewed as single independent entities in a three or four dimensional universe.

It is clear from the complementary manner in which the wave and particle behaved during the dual nature of light controversy that the wave and the particle must be complements. None of the experiments conducted, during the controversy, no matter how sophisticated were able to show the wave nature of light and the particle nature of light in the same place at the same time. During the dual nature of light controversy numerous experiments were purposely conducted, using every conceivable means and scientific principle, trying to trick the particle and the wave to disclose their existence in the same place at the same time. But all the experiments conducted failed to show this, yet one experiment could be conducted showing the wave nature of light, and another experiment could be conducted showing the particle nature of light, but none of the experiments could show both parts of the complement together at the same time.

These results clearly show that light consists of a wave complementary part and a particle complementary part. These complements are space separated complements because any experiment can show the existence of either part of the complement at any time, meaning that both parts of the complements must exist at the same time; and because both parts of the complement exist at the same time they can only exist as space separated complements.

Therefore, both theories are correct, however, they are complementary theories. Both theories originated because the scientists viewed each part of the complement as being separate entities. In so doing, they failed to realize that the complements must abide by the laws of complements. See Laws of Complements in the Appendix.

Therefore, for hundreds of years scientists were trying to see the wave and the particle in the same place at the same time, however, we now know from the two state universe theory that this is impossible. This problem is equivalent to a person trying to look at the upstairs and downstairs apartments at the same time. Or someone trying to see the north pole and the south pole of a magnet in the same place at the same time. Anytime a paradox results in the three or four dimensional universe you can be sure that there are complements involved, in which the observer mistakenly views both parts of the complement as separate entities and then tries to put the two complementary parts together in the same place at the same time.

Heisenberg's uncertainty principle resulted not because of the presence of the observer or because the observer was disturbing the position or the momentum of the particles while trying to measure either part of the complement, rather the uncertainty principle results because of the complementary nature of the universe. Heisenberg's uncertainty principle clearly shows that position and momentum are complementary conditions, which means that both parts of this complement can exist in the two state universe only if they are kept separated in time or space. Therefore, because of the laws of complements, both parts of this complement could never exist in the same place at the same time.

There is no doubt that an observer can affect the environment where an event is taking place which does, indeed, affect the event, such as when an electronic technician is trying to measure a voltage in a circuit, and his meter is drawing current from the circuit. The current being drawn by the meter will affect the measurement. However, if you are trying to see an on light and an off light at the same time, there is nothing that an observer does that changes the fact that these conditions are complementary conditions which cannot exist in the same place at the same time.

A thought experiment showing how an observer can obtain a different result depending on his position in reference to a complementary event is as follows:

From the electrostatic and electromagnetic theory, whenever there is a moving charge there is also a magnetic field associated with it. The magnetic field is energy surrounding the space of the moving charge, and it doesn't exist if the charge stops moving. However, the role of the observer plays an important role in his perception of the existence or non-existence of the magnetic field. For example, if he travels with the moving charge he would not be able to detect the magnetic field, but if he moves far enough away from the charge to observe the magnetic field he would not be able to observe the moving charge.

This thought experiment is the equivalent to what the scientists were trying to do in the wave particle paradox experiment; because if they tried to zero in and observe the particle nature of light they would observe the particle but not the electromagnetic field. By moving away from the particle they would not observe the particle; but they would detect the electromagnetic field or the wave part of the light complement. Both parts of this space separated complement cannot exist or be in the same place at the same time to be observed, because in the two state universe theory, the laws of complements prevents both parts of this complement from existing in such a state, and the most sophisticated experiment or equipment will never be able to detect both parts of this complement in the same place at the same time.

Thus both theories of light are correct because each applies to one or the other complementary parts of light. The difference in the theories is in the time and the space observations from which they are derived. In observing the particle complementary part of light we see the particle described by the corpuscular theory of light; in observing the wave complementary part of light at a different time and at a different place we observe the wave described by the wave theory of light.

Since light is a wave particle space separated complement, we can represent light by combining the wave and the particle complementary parts of the light in one drawing showing them space separated as required by the laws of complements. See Figure 2.

Figure 2. shows an accelerating charged particle with its associated complementary wave. The two complementary parts (the particle and the wave) are shown to be space-separated complements. The figure also shows the wave spreading out in all directions until its radius of curvature becomes a wave front. That is, the radius or distance from the particle is so great that the curvature of the radius appears to approach a straight line. However, since the wave travels in all directions the wave approaches a flat area, rather than just a line. The further away the wave front travels from the particle, the flatter it becomes. This expanding wavefront travels in the manner suggested by Huygens. Therefore, Huygens theory and his model of the wave can also be used to describe the wave complementary part of light.

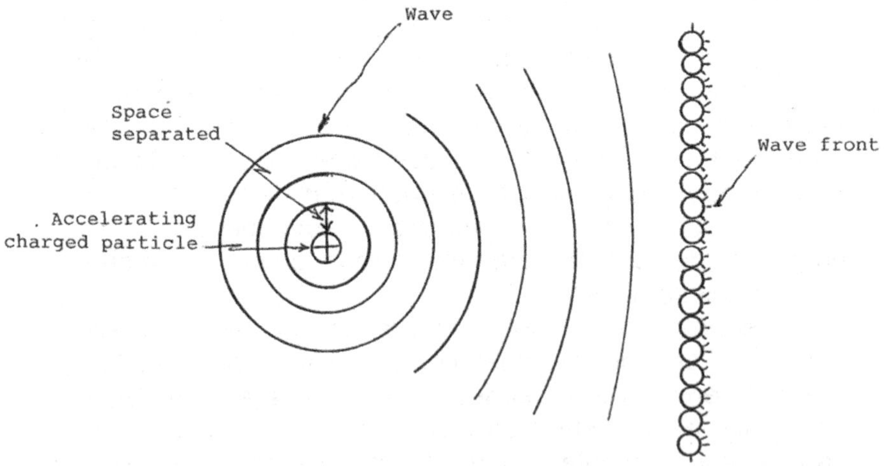


Figure 2.

We are now in a position to examine the two state universe space consisting of infinite numbers of two state points to determine if such a space would have the properties necessary to explain how the wave of Huygens travels through space.

Viewing the aether space from the three or four dimensional time-space theory we find that space contains absolutely nothing, and since it was believed that the aether was a hypothetical medium, described as being jellylike, it is difficult to construct such a medium in a vacuum of space that contains absolutely nothing. But viewing the aether space from the two state universe theory, we know that the vacuum of space is composed of infinite numbers of two state points—a concept different than a point of nothing, and a rather important difference at that, because in a two state point there can be nothing or something existing at each point in space at different times, which could account for the seemingly jellylike nature of the aether, because when the points are in the nothing state they would look like a vacuum of nothing to material bodies traveling through them. However, these same points could change states and act as containers for holding bundles of energy as the energy travels through space from point to point in space and as the points in space changes states from the nothing state to the something or energy state. See Figure 3.

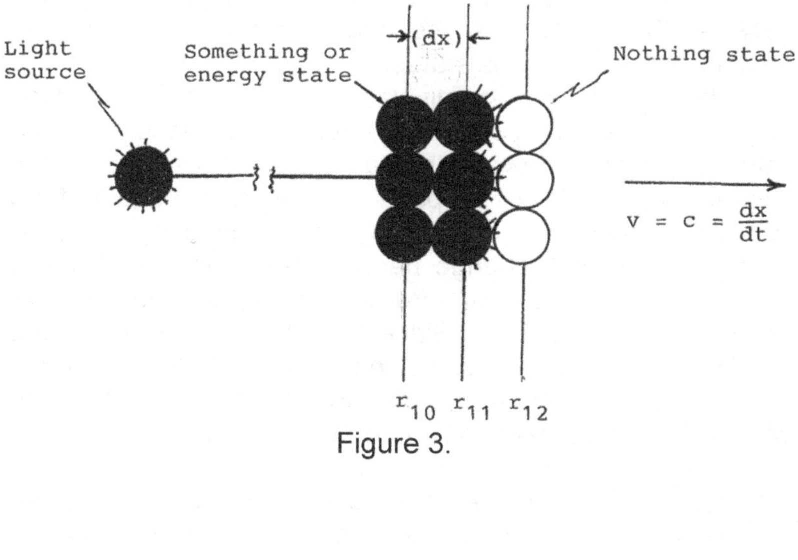

Figure 3.

In Figure 3, we show how a wave, such as the wave described by Huygens, could travel in the space of a two state universe. The figure shows a source of light radiating energy in all directions. The radiating energy travels from point to point in space in the manner described by Huygens. That is, each point in space that receives energy acts as a point source of light. After a period of time, some of the radiated energy reaches the points in space located on the wavefront at $r_{10}$. When the energy reaches these points in space, they change states from the off state to the on, or energy, state. The points at $r_{10}$ are shown in the energy state. The points located on the wavefronts of $r_{10}$, $r_{11}$ and $r_{12}$ are shown greatly enlarged for clarity. After a short time dt (the time it takes the points to change states), the points located on the wavefront at $r_{11}$, receiving the radiated energy from the point sources of light located at $r_{10}$, changes states from the off state to the on state, and these points will also act as point sources of light. The points located at $r_{12}$ did not have enough time to change states; therefore, they are shown in the off or nothing state. The figure also shows that the velocity v at which the light energy travels through the two state universe space is equal to the distance dx divided by the time dt, where dx is the distance between the points in space and dt is the time it takes a point in space to change states.

Velocity is a distance divided by a time, such as miles per hour, feet per second or meters per second. The velocity v is also shown in the figure to be equal to c, where c is the velocity of light in space. Therefore, we see that the velocity of light c is a constant, meaning that it doesn't change. It doesn't change because the velocity of light in space can only travel through space in the manner described in Figure 3. Since the distance between the points in space is a constant and the time it takes for each point in space to change states is also constant, then the velocity c is a constant because its value depends only on the constants of the points in space.

Thus we see that a two state universe space would have the properties of an aether space; because when the two state points are in the nothing state they act as nothing offering no resistance to a material body, such as the planets as they travel through space. Yet these same two state points by changing states would become containers for the energy to travel in as it travels from point to point in the two state universe. These properties would indeed give space its jellylike properties necessary for energy to travel in. These points in space can also account for the quantum nature of energy, because each point in space acts as a container or bundle of energy as the energy travels from point to point through space. And when the energy is received from these points, it would be received only in the fixed amounts of energy contained in these points in space.

We also see that the two state points can account for the constant velocity of light. We can further see that the velocity of light through space would also be constant and independent of the velocity of the source producing it, because if we put a spotlight on a fast moving rocket ship the speed of light would not be dependent upon the speed of the rocket ship because light can only travel through space by going from point to point in space in the manner described above.

Although light travels through space at a velocity that is independent of the velocity of the source, the velocity of light measured by an observer would be dependent on the velocity of the observer, because if the observer is standing still the light would pass the observer with a velocity c of light. But if the observer was traveling with a velocity v towards the oncoming light, the light would pass the observer with a velocity that is equal to the velocity of light c plus the velocity of the observer v. That is, the light would pass by the observer with a speed that is faster than the speed of light in the aether.

If, however, the observer is traveling with a velocity v parallel to and in the same direction as the velocity of light, the light would take a longer time to pass the observer, because this speed would be equal to the speed of light c relative to the aether minus the speed of the observer v relative to the aether.

The model described in Figure 3 can also be used to describe diffraction and interference. However, since diffraction and interference have been fully described in numerous textbooks using the wave theory of Huygens, we will not describe it here.

This chapter clearly shows how easily a paradoxical problem, such as the wave particle paradox, can be solved when viewed from the two state universe theory. In solving this paradox, we were also able to show that:

1. Light travels through the aether space at a velocity equal to the distance between points in space divided by the time it takes the points to change states.

2. Light travels through the aether space at a velocity that is constant and independent of the velocity of the source that produced the light.

3. Although light travels through the aether space at a velocity that is constant and independent of the velocity of the source that produced the light, the velocity of light as measured by an observer would be affected by the velocity of the observer.

# 11.
# The Michelson Morley Experiment

Earlier it was pointed out that the Michelson Morley experiment failed to detect an eather space. The negative result of that experiment had a profound effect on the thinking of a majority of the scientists of that period, changing the way they not only viewed space but also how they viewed the universe.

The thinking about that experiment continues until the present. Because of the great affect that this experiment had on science, we will reexamine the experiment, and then using the laws, postulates and conclusions of the two state universe theory, we will try to determine if there are other reasons that could explain why the experiment failed, other than the reasons given by the scientists of that period.

When the scientists proposed an aether space, as a means for the wave properties of light to travel in, many scientists attempted, theoretically and experimentally, to detect it and determine its properties. However, they were unable to do this. Therefore, they began to believe that it could never be detected.

In l864, James Clerk Maxwell published his treaties on electricity and magnetism wherein he showed mathematically the connection between the electromagnetic theory and light. He believed that he had succeeded in describing the aether and the manner in which light travels through it. He also suggested a way to measure the absolute motion of the earth in the aether, suggesting that it could be done by measuring the speed of light in different directions on the surface of the earth. Then by comparing the differences in the speed of light in the different directions it would be possible to determine the absolute velocity of the earth in the aether. He believed this measurement was possible because the earth travels through the aether with a speed of at least 30 kilometers per second as it orbits the sun. However, he didn't believe that this measurement could be made, because the sensitivity of the equipment needed to make these measurements was not available at that time.

In 1881 Albert A. Michelson (1852-1931), a U.S. Navy officer, after being granted a leave of absence from the Navy to study in Europe, devised an ingenious experiment, along the lines suggested by Maxwell, to measure the absolute velocity of the earth. If successful the experiment would not only measure the velocity of the earth in the aether but it would also establish the existence of the elusive aether.

Michelson's experiment is often described by the analogy of a swimmer swimming back and forth across a stream and then swimming an equal distance up and down the same stream. Mathematically it can be shown that it takes longer to swim upstream and downstream than back and forth across the stream.

Based on this analogy, Michelson designed an experiment using an interferometer which he believed would be capable of measuring the absolute velocity of the earth in the aether. This experiment was carried out in the laboratory of Herman von Helmholtz at the University of Berlin. The first interferometer made by Michelson was crude, not too sensitive and difficult to operate.

Later he had an instrument maker make a more sensitive interferometer. It was installed and operated at the Astrophysical Observatory in Potsdam in Berlin. The optical parts were mounted on two crossbeams, but unfortunately they were not rigid enough to give reliable results. A representation of this interferometer is shown in Figure 4.

The interferometer uses a light beam which is split, by means of a half-silvered mirror, into two paths $t_1$ and $t_2$. One is directed 90 degrees to the other. By means of mirrors the beams are reflected from a light source through the two paths to and combined at an eyepiece, where interference occurs if the time for each path differs.

If the two paths are equal, then the two rays arrive at the eyepiece in phase, meaning that the positive and negative peaks of the light waves from each beam occur at the same time. These two waves add at the eyepiece, which results in a brighter light. When the waves add and give a brighter light this is called constructive interference.

Figure 4

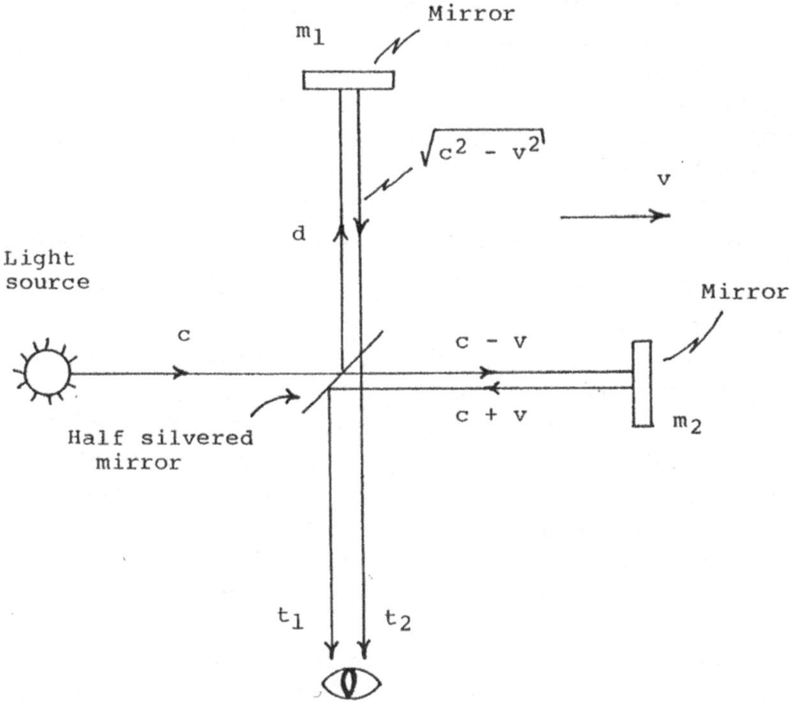

If the difference in the paths are such that the waves arrive at the eyepiece 180 degrees out of phase, meaning that the positive and negative peaks of each wave arrive in opposite directions; then when the waves are added at the eyepiece they will tend to cancel each other resulting in a darker light. This type of interference is called destructive interference. If one of the mirrors is tilted slightly, a number of light and dark fringe patterns result. If the difference between the two paths change, the light and dark fringes move past a fixed point and are counted, indicating the difference between the two paths.

If the time it takes for each path to reach the eyepiece is calculated, we find that the equations derived for the interferometer are exactly the same as the time equations calculated for the swimmer swimming in the stream. Therefore, it takes longer for the $t_1$ beam of light to arrive at the eyepiece than for the $t_2$ beam of light. This is because the $t_1$ path is the path that the $t_1$ beam travels upstream and downstream in the aether wind.

Because the arms of this interferometer built for Michelson were crude and not sturdy enough to prevent vibrations, the interference fringes were not steady, and they were difficult to read. Therefore, this experiment, conducted at Potsdam, was considered by Michelson to be unreliable. However, he published the results of the experiment, which showed that there was little or no interference. This conclusion meant that the hypothesis that there is an aether space is not correct.

Michelson was disappointed by these results, and he did not care to pursue the aether space experiment any further. After reading about Michelson's failure, some scientists believed that the concept of an aether space should be discarded; however, other scientists didn't believe that the concept should be discarded because the results were inconclusive. Later in 1881 Michelson resigned from the Navy to become a professor of physics at the Case School of Applied Science in Cleveland. The school is now known as the Case Institute of Technology.

While at the Case School of Applied Science, he met Edward W. Morley (1838-1923), a chemist, mathematician and physicist, who had an interest in the aether space problem. Morley was from the Western Reserve University, located near the Case School of Applied Science.

Michelson and Morley became good friends, and in 1884 they both attended a lecture given by Lord Kelvin in Baltimore, which was partly about the luminiferous aether. It was during this trip that Lord Kelvin and Lord Rayleigh tried to encourage Michelson to continue his aether space experiments.

When Michelson and Morley returned home to Cleveland, they were convinced that the experiment should be given another trial using more sophisticated equipment. At Morley's suggestion, a new interferometer was constructed containing a sandstone slab table about five foot square mounted on a pool of mercury, to prevent the vibrations that occurred at Potsdam. The table contained a number of mirrors to give a greater path length for the light ray to travel. This arrangement gave this interferometer a sensitivity that would cause four-tenths of a fringe shift when the instrument was rotated 90 degrees in the 30 kilometer aether drift. This was much more sensitive than the instrument used at Potsdam, which gave only four-hundredths of an interference fringe when the instrument was rotated the 90 degrees in the aether wind.

The equipment was originally being constructed in the main building of the Case School of Applied Science. However, a fire broke out, but fortunately most of the equipment was removed from the fire. After the fire the equipment was set up at the Western Reserve University.

The experiment was conducted in July, 1887. The equipment, when completed, was sensitive enough to detect the speed of the earth to approximately 1 km/sec. When the experiment was finally performed, Michelson and Morley were disappointed because the result showed little or no fringe shifts.

This experiment was another failure for Michelson in his attempt to detect the aether; and aside from being a great disappointment to Michelson and Morley, it had a profound affect on the thinking of a majority of the scientists, who began to explore all sorts of reasons for the failure.

George F. Fitzgerald (1851-1901) proposed that the interferometer flattened in the direction of the velocity, due to the pressure applied to the interferometer as it moved through space. What he was proposing was that an object or the equipment shrinks in the direction of the velocity, and in just the right amount to cause the beams in the two paths to be equal. This shrinkage with velocity would cause the interferometer to record no interference.

Expanding on that proposal, H. A. Lorentz (1853-1928) expressed this shrinkage in the direction of the velocity in his famous formula which states that the rest length of an object shrinks by the value of the square root of $(1-v^2/c^2)$ multiplied times the objects rest length when the object moves with a velocity v. This theory is known as the Lorentz-Fitzgerald contraction theory.

Einstein learned of the Michelson Morley experiment from his friend Lorentz, which led indirectly to his special theory or relativity. In his paper titled "On the Electrodynamics of Moving Bodies" Einstein suggested " . . . the phenomena of electrodynamics as well as of mechanics possess no properties corresponding to the idea of absolute rest. . . ." If there were no aether or absolute rest; then everything would have to be measured relative to some other moving object. If the aether had been found to exist, it could have been used as a fixed reference point for all measurements.

The negative results of the Michelson Morley experiment profoundly affected the direction and the thinking of scientists who were completely puzzled by the failure. Michelson repeated the experiment again in 1904 with even more precise equipment. The results were even more negative and conclusive than the experiment performed by Michelson and Morley at Western Reserve.

In the next chapter, after analyzing the experiment from the two state universe theory, we will be able to give reasons why the experiment failed.

# 12.
# Why the Michelson Morley Experiment Failed

The failure of the Michelson Morley experiment was a turning point in the history of science, and it caused a change in the way scientists viewed the universe. Many bizarre hypothesis and ideas were proposed as the scientists tried to explain this unexpected result. The failure of that experiment still affects the way scientists think about the function of the universe, and it will continue to do so for many years to come.

Because of the unexpected result of the experiment and some of the bizarre and weird hypothesis still being proposed today by the scientists, we will reexamine the Michelson Morley experiment from the two state universe theory to determine the reason why the experiment failed and if, indeed, it failed because there is no aether.

Let's examine in more detail the paths of the light rays as they travel from the light source to the eyepiece in Figure 4. These paths appear reasonable on first inspection, but if we take a closer look, we discover there are at least three possible paths that can be followed from the light source to the eye as shown in Figure 5.

In Figure 5 (a), we see a path taken by the light if we assume the experiment was ideal and the equipment was constructed with perfect mirrors (perfectly flat and smooth).

Figure 5.

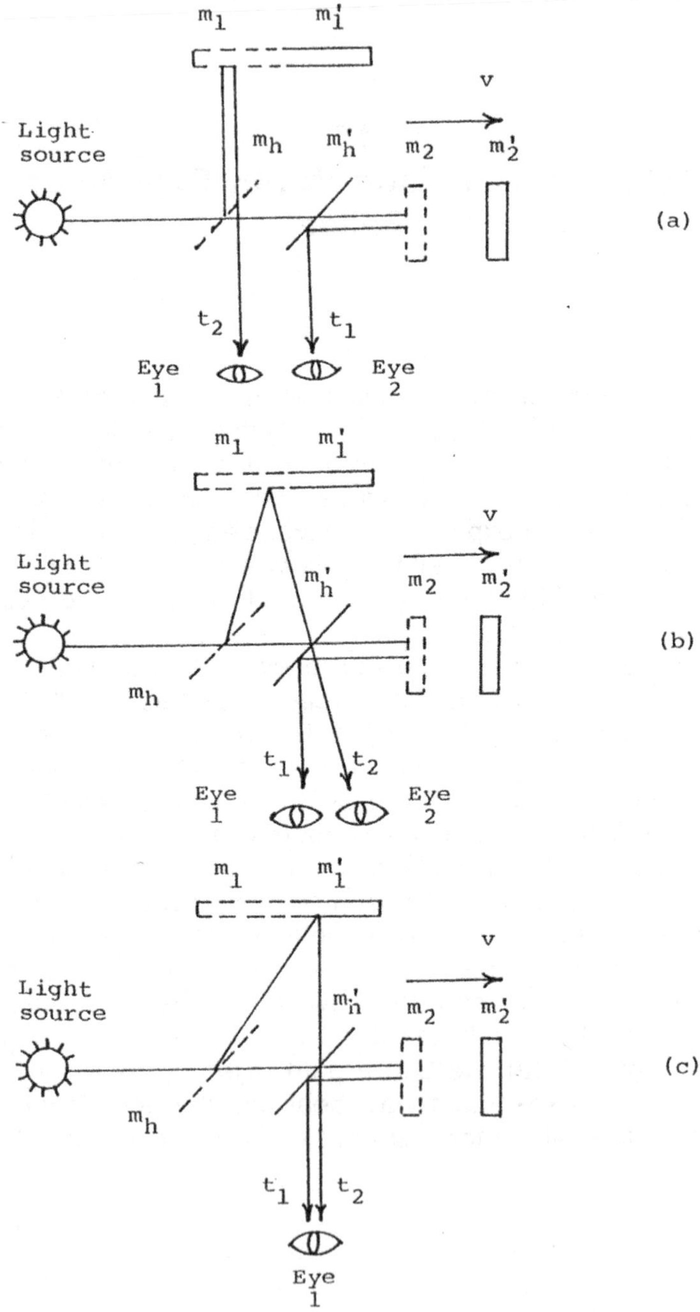

m1
m'1
v
Light source
mh
m'h
m2
m'2
(a)
t2
t1
Eye 1
Eye 2
m1
m'1
v
Light source
m'h
m2
m'2
(b)
mh
t1
t2
Eye 1
Eye 2
m1
m'1
v
Light source
m'h
m2
m'2
(c)
mh
t1
t2
Eye 1

The path starts from the light source; travels to $m_h$; and because the angle of incidence equals the angle of reflection, the light is reflected straight up in the aether to $m_1$. On reaching $m_1$ it is reflected straight down to eye 1. This is the $t_2$ path. Since $m_h$ is half silvered, it passes half of the oncoming light through to $m_2$. Because the instrument is traveling through the aether with the velocity v, the mirrors all move in the direction of the velocity. At some time after passing through $m_h$, the light ray meets $m_2$ and is reflected back to the left in the direction towards $m_h$ where it meets it. The ray is then reflected straight down to the eye 2. And, as can be seen, $m_2$ continues to move to $m'_2$ after the ray is reflected from it. This is the $t_1$ path. From the drawing we can clearly see that the eye of the experimenter would have to be in two places at the same time in order for the experimenter to see both rays. But from Law 4 of complements (something that exists cannot be in two different places at the same time), we see that the eye can't be in two places at the same time. Therefore, this experiment would fail because there wouldn't be any interference, as the viewer would only see the light of the $t_1$ path. He wouldn't see the two rays at the same time in the same place in his eye, the condition needed for interference.

Figure 5(b) shows the path used in the Michelson Morley experiment on which they based their calculations. In this figure the light from the source travels to the right where it meets $m_h$. It is then reflected at some angle up and to the right, to $m_1$. We assume the mirror has an imperfect surface to reflect it in the direction indicated, otherwise the ray would reflect straight up in the aether. This would occur because if the ray is reflected from a perfect mirror, the angle of incidence would equal the angle of reflection and the ray would then be directed upwards at an angle of 90 degrees from the on coming light source which is directed horizontally to the right. Once the light ray is reflected, it travels in the aether at a speed c and in the direction it was reflected in, and the motion of the equipment would not alter it. That is, a light ray travels in the aether at a constant rate c and in a direction not altered by the motion of any vehicle that may be near, carrying or passing through the aether. See conclusion 23 in the appendix.

If the mirror is not perfectly smooth, or even if it is slightly tilted, the light ray could then take the path indicated in the drawing, which is up and to the right. The ray continues until it reaches $m_1$ where it is again reflected down and to the right where it continues through the aether to eye 2. This is the $t_2$ path. Also because of the half silvered mirror, part of the light passes through to mirror $m_2$ where it is reflected back and to the left until it reaches the half silvered mirror, which has arrived at $m'_h$. The ray is then reflected straight down to eye 1. This is the $t_1$ path. $m_2$ continues to move to $m'_2$ after the ray was reflected at point $m_2$. From this path we see that the eye would also have to be in two places at the same time. Therefore, the experiment would fail to show interference. The experimenter could only see the ray of $t_1$ because it is reflected straight down to the eyepiece. Unable to see $t_2$ at the same time in the same place as $t_1$, there would be no interference.

In examining Figure 5(c), we see one more possible path. This path is possible because the light, which splits into two parts, could travel from the source to the eye and arrive in the same place, so that the eye could observe both paths at the same time. Examining this path, we see that a beam leaves the light source and travels horizontally and to the right until it meets $m_h$ splitting the beam at this point. One part travels at an angle, up and to the right—this angle is possible if the mirror is not perfect; the imperfect surface could then reflect the light in the direction shown. On reaching $m'_1$, the light could reflect the ray straight down to the eye 1. This path is also possible, if there are imperfections on the surface of the mirror, which would reflect the light along these paths. This is what happens because mirrors are not perfect. This same path could also occur during the adjustment of the interferometer. When the operator adjusts the mirrors to produce the interference patterns; the mirrors could easily be tilted at just the right angles to produce this path. This is probably what occurs in an interferometer, because the adjustment is made merely to produce the interference patterns not to insure a particular path.

In the second path, after passing through $m_h$ the light travels horizontally and to the right until it reaches $m_2$; it is then reflected straight back horizontally and to the left, until it meets the half silvered mirror, which has now moved through the aether and arrived at $m'_h$. The ray is then reflected straight down to the eye 1, which is in the same place at the same time as the ray of $t_2$. This second path is the ray of $t_1$. $m_2$ has moved in the aether to $m'_2$, after reflecting the light ray at $m_2$. This path—from the light source to the eye—is possible, and is such that both paths are capable of entering one eye of the observer parallel to each other in the same place. If the time of the paths were different, we would be able to detect this difference by the interference produced.

If we calculate the length of the paths for the two rays and the time it takes for the light rays to travel the length of each of the two paths in Figure 5(c), we find, there are two solutions for this problem.

One of the solutions shows that the time $t_2$ is equal to zero. This solution is unexpected because it suggests that the speed of light could be infinite or instantaneous. We know, however, that the wave complementary part of light travels at the value c, which is the speed of light in the aether. We have seen in Chapter 10 that the wave part of light can only travel through space with a velocity that is dependent on the distance between the points in space divided by the time it takes the points in space to change states. There is, however, nothing in the two state universe theory that limits the velocity of the particle nature of light. This solution of zero time would give a correct solution to this problem if the particle complement of light traveled instantaneously in both paths from the light source to the eyepiece. In that event the particle complement of light from both paths would arrive at the eyepiece at the same time and produce no interference. More time, however, would be needed to more fully explore this unexpected result of zero time for $t_2$.

The second solution to the paths of Figure 5(c), shows that the time $t_1$ and $t_2$ are equal. This means that the path lengths are the same. Therefore, the rays from both paths would arrive at the eyepiece at the same time, and the waves of both rays would arrive at the eyepiece in phase, resulting in no interference.

This is indeed a very surprising and unexpected result! The three possible paths lead to the same conclusion that there can be no interference for any of the three paths. This is completely contrary to the calculations of Michelson and Morley. Based on the theory of the two state universe, we have arrived at the surprising and unexpected result that there can't be any and there shouldn't be any interference in the experiment designed by the two scientists. What does this mean? It means that the length of the interferometer doesn't change in the direction of the velocity, nor does it change in any direction with velocity. If the length changes, as suggested by Lorentz-Fitzgerald, the equations suggest there would have to be an interference pattern shown between the paths of Figure 5(c), because these paths are certainly possible. Since no interference was detected, there can't be any contraction of the interferometer.

How does this result affect the theory that mass increases with velocity? It would appear that this result, of no contraction in the direction of the velocity, should have some effect on the theory that mass increases with velocity. We will examine this question later.

The unexpected and startling conclusion proved that the results of the Michelson Morley experiment were not negative; they were as they should have been. Because of this, we must come to the following conclusion:

24. Velocity does not have any effect on the contraction of an object; an object does not contract in the direction of the velocity.

This startling conclusion clearly shows the effectiveness of the two state universe theory in solving complex problems.

The interferometer, therefore, can't be used to detect the absolute velocity of the earth in the aether. This result, however, does not rule out an aether space nor does it rule out other means of detecting an aether space. In fact it clearly shows that there must be an aether space, because the problem was solved using the concept of the interferometer moving in the aether, and since the calculations took the aether space into consideration and the results of the Michelson Morley experiment were in agreement with these calculations, there must be an aether space; otherwise, the experiment would have shown interference. We will examine the aether space in more detail in the next chapter.

Thus we cannot rule out an aether space, and if we can't rule out the existence of an aether space, we would also have to conclude that:

25. Time does not decrease with an increase in the velocity, nor does it change with velocity.

With the establishment of the possibility of an aether space, we would also have to reestablish that time is an invariant. This is logical because in the two state universe theory time is nothing. Therefore, how could nothing, which has no characteristics, change with velocity?

These two conclusions are included in the conclusions located in the Appendix.

# 13.
# Is there an Aether Space

We have seen that the Michelson Morley experiment was not the complete failure believed by the two scientists, because the results were exactly as they should have been. Our calculations showed that the interferometer should not produce interference, which is exactly what occurred. Expecting interference, and finding none, Michelson and Morley concluded that their experiment was a failure.

Lorentz and Fitzgerald reasoned that it failed because the interferometer contracted by just the right amount in the direction of the velocity, causing the two paths to be equal, and if the two paths are equal, there wouldn't be any interference. However, since our analysis shows that the experiment didn't fail, the contraction theory could only be plausible if there was interference.

Although the experiment did what it was supposed to do, it failed to detect the absolute velocity of the earth in the aether. Does the results of this experiment mean that we can never measure the absolute velocity of the earth in the aether?

In view of what we now know about the experiment, the answer is no. Our calculations were based on the existence of an aether space, and since the results were exactly as predicted, the results of the experiment does suggest the existence of an aether space. And if an aether space exists we would then have an absolute reference point for measuring the velocity of the earth in the aether.

In my book titled, The Two State Universe, I suggested a way to measure the velocity of the earth in the aether, which would also show the existence of the aether space.

In the book I suggested using a large rotatable platform, such as an aircraft carrier on which the experiment could be conducted. A laser would be mounted at one end of the carrier, and a recorder would be mounted some distance from the laser. If the aircraft carrier is rotated to face in the direction of the velocity of the earth in the aether and the laser light is positioned to also face in that same direction, then when the laser light is turned on, the ray of light would leave the laser and travel from point to point in the aether until it reaches the recorder. The point where the laser light lands on the recorder would be a reference point. Then the carrier is rotated one-quarter turn (90 degrees), and the laser is again turned on. Now, when the laser light leaves the laser it travels from point to point in the aether until it again reaches the recorder. However, with the laser light traveling at 90 degrees to the velocity of the earth in the aether, the carrier, the laser and the recorder will all be traveling in the direction of the velocity of the earth in the aether, and at the same time the laser ray travels from point to point in the aether, which is fixed and does not move. When the laser light reaches the recorder, the recorder would have moved a small distance in the direction of the velocity of the earth in the aether, and the spot where the light lands on the recorder will then be that same small distance from the reference point.

The amount of the deflection depends on the velocity of the earth in the aether, the distance of the recorder from the laser and the velocity of the laser light in the aether. Using a value of 30 kilometers per second for the velocity of the earth in the aether, 500 feet for the distance of the recorder from the laser and $3 \times 10^8$ meters per second for the velocity of the laser light in the aether, we get a deflection of .6 inches to the right of the reference. Although small this deflection is measurable. If the carrier is then rotated to 180 degrees, 270 degrees and 360 degrees back to its original direction of zero degrees facing in the direction of the velocity, we will get a zero deflection from the reference when the carrier is at 180 degrees, a deflection of .6 inches to the left of the reference when the carrier is at 270 degrees and zero deflection from the reference when the carrier is at

360 degrees. This is because at 360 degrees the carrier would then be back to its original position of zero degrees where it started.

In this experiment we pointed out that if the laser point returns to the same reference point when the carrier is rotated 360 degrees (i.e. when the carrier makes one complete cycle and returns back to its original position), we could be certain that the carrier is not being distorted during its 360 degrees of rotation. We also see that the deflection at 90 degrees is .6 inches to the right of the reference and it is .6 inches to the left of the reference at 270 degrees giving a total deflection of 1.2 inches. This deflection can easily be measured.

We also suggested using mirrors to lengthen the path of the laser ray 5 times, giving a total deflection of 6 inches. And if we add one more mirror to lengthen the path 6 times, the recorder would be in line with the laser light making the recorded dot less sensitive to deflections or warping of the carrier. This arrangement would not only make the recorded dot less sensitive to the motions of the carrier but it would also result in a total deflection of 7.2 inches. This is a considerable deflection, which can be used to easily calculate the velocity of the earth in the ether, and in so doing we would establish the existence of the aether.

Although this experiment can easily measure the velocity of the earth in the aether, it is a difficult experiment to do because it requires the use of an aircraft carrier, a calm sea or lake to conduct the experiment in and there would be some doubts about whether or not the aircraft carrier would warp or bend during its 360 degree maneuver needed to do the experiment. Therefore, an easier experiment using the earth as the moving platform will now be described.

The earth is an ideal platform to use for this experiment because the earth rotates 90 degrees every 6 hours, 180 degrees every 12 hours, 270 degrees every 18 hours and 360 degrees every day. This occurs because the earth rotates counter clockwise on its axis making one revolution (360 degrees) in a day. See Figure 6.

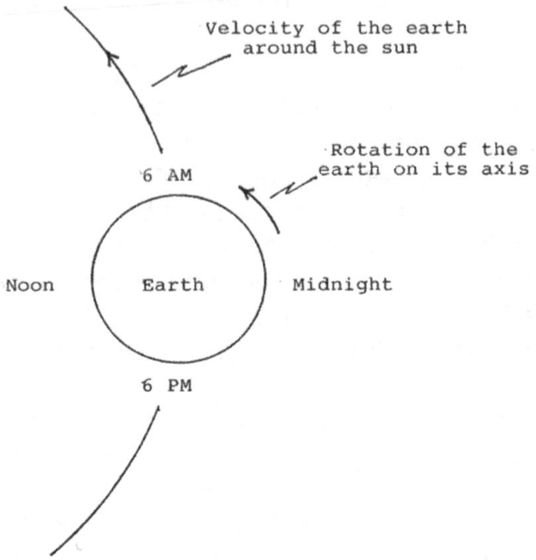


Figure 6.

Figure 6 illustrates the motion of the earth as it travels around the sun. We see from the illustration that the earth travels around the sun in a counter clockwise direction. It takes one year for the earth to complete one cycle around the sun. The velocity of the earth as it travels around the sun is approximately 30 kilometers per second.

The illustration also shows the earth rotating counter clockwise on its axis. It takes one day for the earth to make one revolution on its axis. We also see from the drawing that at 6 AM the earth faces in the direction of the motion of the earth around the sun; and at noon, 6 hours later, the earth has rotated one quarter turn from the 6 AM position, which means that at noon the earth faces in a direction that is 90 degrees to the 6 AM direction. At 6 PM the earth is facing in a direction that is 180 degrees to the 6 AM position and at midnight the earth faces in a direction that is 270 degrees from the 6 AM position (the 270 degree position is the same as a minus 90 degree position).

Continuing in a counter clockwise direction we see that at 6 AM one day later we will have arrived at a position that is again facing in the direction of the velocity of the earth around the sun. With this information we can now describe a simple experiment that could be used to determine the velocity of the earth in the aether and also show the existence of the elusive aether.

To conduct the experiment, we must securely mount a laser to a point on the earth facing the laser in the northerly direction; and some distance from the laser we must securely mount a recorder R to the earth in such a way that the recorder is stationary. The recorder is positioned north of the laser light so that when the laser is turned on it will indicate by an illuminated spot on the recorder where the laser light lands on the recorder. The recorder could be something as simple as a sheet of white paper mounted on a piece of plywood that is securely anchored to the earth in such a manner that it will not move during the experiment. See Figure 7.

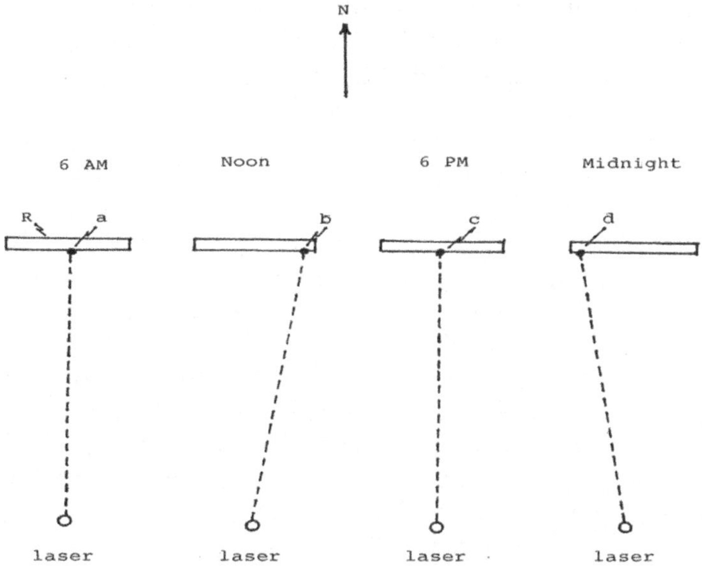


Figure 7.

In Figure 7 we see that if the laser light is turned on at 6 AM, the laser light will travel from point to point in the fixed aether space, and at the same time the recorder R and the laser will be moving in the direction that the earth travels in the aether. At 6 AM this direction is straight up. Therefore, the laser will land at point a which is straight ahead; however, because of the motion of the earth in the aether the a point will land a short distance in the down direction. At noon when we turn on the laser light, the ray of light from the laser travels from point to point in a straight line in the fixed aether; however, at noon the laser (not the light that left the laser) and the recorder is moving with the earth in the direction the earth travels in the aether. At noon this direction is to the left. With the laser ray traveling in the aether that doesn't move then the laser ray will land at b, which is a slight distance to the right. This occurs because if the recorder is moving to the left and the laser ray is not moving with the earth, the laser ray would appear to be moving to the right as it approaches and lands on the recorder.

At 6 PM the laser light travels from point to point in the fixed aether and lands at point c. Point c is in line with point a, however, it moves in a slightly upward direction. This is because at 6 PM the earth with the recorder travels straight down while the laser light ray travels straight towards the recorder in the fixed aether, which doesn't move. Therefore, the light ray strikes the recorder in a slightly up direction. And at midnight, the laser light ray again travels straight towards the recorder; however, it lands to the left at d. This occurs because as the earth and the recorder moves in the direction of the velocity of the earth in the aether, the laser ray travels from point to point in the fixed aether, which doesn't move. The direction of the earth in the aether at midnight is to the right and, since the aether doesn't move, the light ray lands slightly to the left on the recorder.

The amount of the deflection in each of these cases depends upon the distance that the recorder is located from the laser and the velocity of the earth in the aether. If we assume that the velocity of the earth as it travels in the aether is at least 30 kilometers per second (the velocity of the earth as it travels around the sun), and we set the distance between the recorder and the laser to 500 feet, we get a deflection of .6 inches in each of the four cases described. See Figure 8.

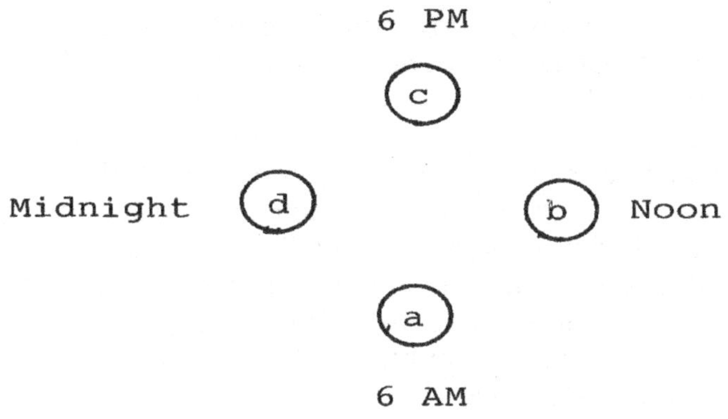


Figure 8.

In Figure 8 we show what the recorder would indicate if we looked straight ahead at that area of the recorder where the light ray strikes it. With the deflections of .6 inches described above, the total deflection indicated on the recorder would be 1.2 inches between 6 AM and 6 PM. It would also be the same distance between noon and midnight. If the distance to the recorder were doubled to 1,000 feet then the total deflection in each case would be 2.4 inches.

If we get a deflection of at least 1.2 inches, at 500 feet, this would clearly show that the earth travels in the aether with a velocity of at least 30 kilometers per second, and it would also confirm the existence of the aether.

This experiment could also be conducted with the laser facing in an easterly direction; in fact you should be able to detect the deflections if we conduct the experiment with the laser facing in any direction—north, south east or west.

With the laser facing in the easterly direction, we would find that the noon and midnight points would land midway between the 6 AM and the 6 PM position. This occurs because at noon the laser is facing east which is in a direction that is directly opposite to the direction the earth is traveling in the aether. This results in no deflection of the laser ray. Therefore, it would land midway between the 6

AM and the 6 PM point, and at midnight the laser would be facing directly in line with the velocity of the earth in the aether, resulting in no deflection. This causes the laser ray to land also between the 6 AM and the 6 PM marks. At 6 AM the laser is pointed 90 degrees to the direction of the velocity of the earth in the aether. This results in the laser light being deflected in a down direction. And finally at 6 PM the laser would be facing 270 degrees or minus 90 degrees from the direction of the earth in the aether, resulting in the laser light striking the recorder in a slightly up direction.

This experiment is a rather simple experiment to conduct; and if it gives the results described above, there could then be no doubt about the existence of an aether space. The experiment would also give an experimental value for the absolute velocity of the earth in the aether.

# 14.
# Results of the Laser Experiment

After writing the last chapter in which I described the experiment that uses the earth as a moving platform, I decided to conduct the experiment myself, because the experiment appeared to be simple and straightforward. I had available a laser pointer which, when I purchased it, was advertised to travel l,500 feet. I found a large piece of plywood in my basement, which was used as part of the recorder.

The first time I conducted the experiment, I mounted the laser on the wooden floor of our outside porch. I mounted it in such a position that when the laser was turned on the laser ray would land on the plywood, which was mounted on a large tree trunk that was located about 125 feet from the point where I mounted the laser.

The spot I selected for mounting the laser was within about one and a half foot from a concrete column that was anchored in the ground as a support for the floor of the porch.

I turned on the laser and noted on the plywood where the laser ray landed. I then attached a sheet of typing paper to the plywood in such a position that the point where the laser ray landed would be approximately in the center of the typing paper.

The first experiment was conducted in the early summer of 1998. Since the experiment was conducted during the time of the year that daylight saving time is used, I decided that I would take the readings at 7:00 AM, 1:00 PM, 7:00 PM, and 1:00 AM. The reason I selected these hours was because I wanted to record the results using the standard time of 6:00 AM in the morning when the earth at that time faced in the direction of the velocity of the earth around the sun. I also wanted to record the standard time readings of 12:00 noon, 6:00 PM in the evening and 12:00 o'clock midnight.

When I took my first few measurements, I was in for a surprise, because I found that during the 24 hours that is needed to conduct the experiment there were all kinds of expansions and warping of the porch floor which caused the laser dot, that was recorded on the typing paper, to move haphazardly all over the paper. I tried for a number of weeks to find a suitable way to mount the laser to avoid the movements. Eventually I begin to get some readings that appeared to reveal some kind of a pattern that could reasonably be expected for the experiment, but just when the experiment seemed to be operating as it should, the laser would shift and the laser ray dot would move wildly to another spot which obviously showed that the laser mounting was shifting.

I then tried using a cinder block, which I buried half way in the ground. The cinder block was positioned 134 feet from the recorder. I mounted a three-quarter inch piece of pinewood on the top of the cinder block, and then I mounted the laser pointer on the piece of pinewood. The laser was mounted pointing in a southerly direction.

I conducted the experiment on this platform for another few weeks with little or no success. I tried many different ways to prevent the piece of wood and the laser from shifting during the experiment, but no matter what I did, I continued to get shifting and warping of the board which caused false and inconclusive results.

I eventually decided that the laser would have to be mounted directly on the cinder block to prevent any warping effects being caused by the wood. So I bought some contact cement and attached the laser pointer directly to the cinder block with the contact cement.

Then at 7:00 AM, 1:00 PM, 7:00 PM on September 12, l998 and at 1:00 AM on September 13, l998, I conducted my first successful experiment. The results of the experiment showed a pattern of four dots. The 7:00 AM dot was up, the

1:00 PM dot was to the right, the 7:00 PM dot was down and the 1:00 AM dot was to the left. In order to verify this pattern of four dots, I decided to repeat the experiment on the 13th and the 14th of September. The four-dot pattern repeated itself exactly, with each of the four dots landing on each of the previous experimental pattern of four dots.

I continued the experiment for another two days on the 14th and the 15th of September, using another sheet of typing paper. The results were the same. I was happy with the results, not only because the pattern repeated itself exactly but also because of the information that was being disclosed about the velocity of the earth in the aether by the pattern of the dots.

I decided, because of the results I was getting that I should repeat the experiment again around September 23rd, which would show results during the fall equinox. Figure 9 shows a drawing illustrating the results of that experiment. I begin the experiment at 7:00 PM on September 23rd and continued the readings until 1:00 PM on September 25th.

The results of these experiments were surprising and better than I expected because they disclosed a resultant velocity consisting of the velocity of the earth around the sun, the velocity of the sun around the center of our galaxy and the velocity of the galaxy as it travels in the universe. The results, however, shows only the resultant velocity in the plain that the earth makes as it travels around the sun. It doesn't show the velocity in a different plain, such as a plain located 90 degrees to the plane of the earth as it travels around the sun. Also the resultant velocity represents the velocity of the earth only at the time the experiment was conducted. This is because the resultant velocity is dependent upon the velocity of the earth around the sun, the velocity of the sun around the galaxy and the velocity of the galaxy in the universe, and because the velocity of the earth around the sun is constantly changing, the resultant velocity is constantly changing.

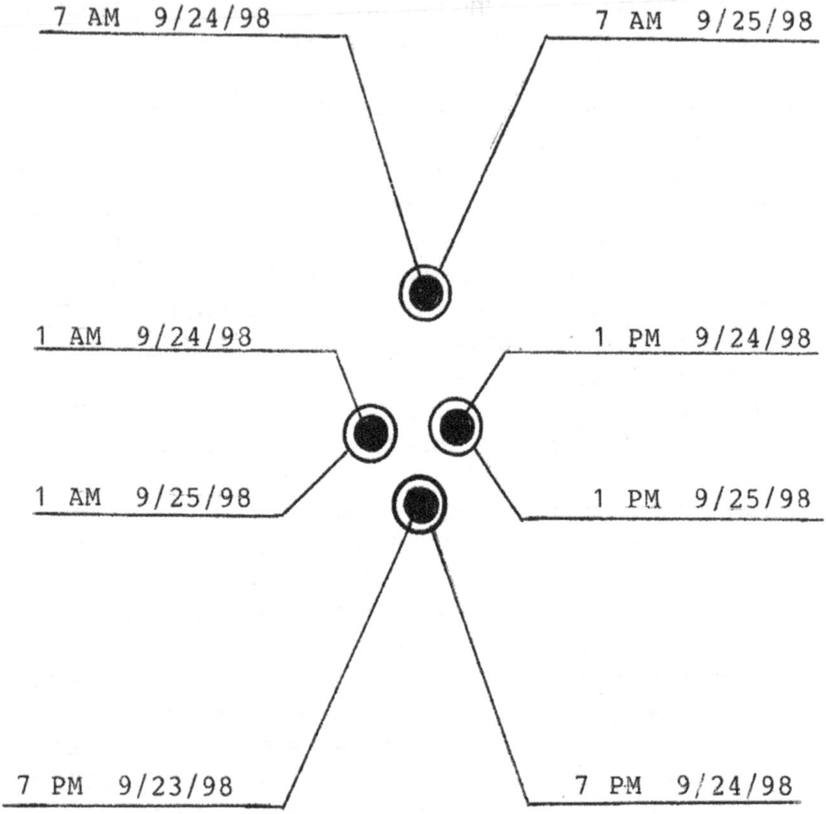


Figure 9.

Figure 9 shows that the resultant velocity of the sun as it travels counter clockwise around the center of the galaxy and the velocity of the center of the galaxy as it travels in the universe is in a direction that is opposite to the velocity of the earth as it travels around the sun, and this resultant velocity is much greater than the velocity of the earth around the sun. We know this because the recorded dots are all in a direction that is opposite to the direction they would be in if there was only the velocity of the earth as it travels around the sun involved. If we only had the velocity of the earth around the sun then the dots shown recorded in Figure 9 would show the 7:00 AM dots and the 7:00 PM dots in reverse positions, and the 1:00 AM dots and the 1:00 PM dots would also be in opposite positions.

If we calculate the approximate resultant velocity in the plane suggested in this experiment, we find the resultant velocity of the earth in this plane to be about 58 kilometers per second in a direction straight out from the 7:00 PM position of the earth. This calculation includes a correction for the approximate 42-degree latitude position of the laser and the approximate 23.5-degree tilt of the axis of the earth.

If we add to this velocity of 58 kilometers per second the approximate 30 kilometer per second velocity of the earth as it travels around the sun, we get a velocity of 88 kilometers per second for the combined velocity of the sun as it travels around the center of the galaxy added to the velocity of the galaxy in the universe. We know that the sun is traveling around the center of the galaxy faster than this value of 88 kilometers per second; therefore, we can conclude that the center of the galaxy must be traveling, at that instant, with a velocity that is in the same direction as the velocity of the earth as it travels around the sun.

If we repeat this experiment again on the first day of spring, we should expect to find that the earth would be traveling with the velocity of the sun plus its velocity as it travels around the sun. Therefore, the experiment should indicate a velocity at that time to be the velocity of the sun added to the velocity of the earth around the sun which would give a velocity of approximately 118 kilometers per second.

Although this experiment was crude, and it is unable to show the absolute velocity of the earth in the aether, it is of a sufficient accuracy to show clearly that there is an aether space, and it also gives a rough approximation of the resultant velocity of the earth as it travels in the aether, but only in the plane that the earth makes as it travels around the sun.

This crude experiment, which gives only an approximate result of the earth's velocity, can easily be improved on by conducting the experiment indoors in a large building, so that the temperature changes could be controlled sufficiently to reduce the effects of warping, and the laser used should be one that has a more concentrated and focused beam than the laser that was used in this experiment to allow for a more accurate recording of the laser ray dots.

This experiment could be conducted on the first day of winter, the first day of spring and the first day of summer to obtain even more information about the velocity of the earth in the aether. For example, on the first day of winter the velocity of the earth around the galaxy should be perpendicular to the direction of the velocity of the earth around the sun. At that time the value of the velocity of the earth around the sun could be calculated. In this experiment we were only able to calculate the approximate resultant velocity of the earth around the sun, the velocity of the sun around the center of the galaxy and the velocity of the center of the galaxy in the universe.

The experiment described in the last chapter using the aircraft carrier could also be used to get an even more accurate resultant velocity, because that experiment could be conducted over a shorter period of time than the 24 hour time required for the earth experiment, and there would not be the errors involved due to the curvature of the earth and the tilt of the earth. Further, the aircraft carrier experiment can be greatly simplified, because we now know that the experiment will indicate a resultant velocity that is much greater than the velocity of the earth around the sun. Knowing this, it would not be necessary to use the mirrors described, and the distance from the laser to the recorder could be decreased. Using a distance of about 300 feet from the laser to the recorder should give satisfactory results. This means that the experiment could be conducted on a ship that is much smaller than an aircraft carrier, and an experiment conducted on a smaller ship will greatly reduce the possibility of errors due to the twisting and warping of the ship.

Thus we find from this experiment that, although we were unable to calculate an accurate value for the resultant velocity of the earth in the aether, we were able to clearly show that there is an aether space.

Knowing that there is an aether space means that there is an absolute reference frame. This finding, if verified by the scientific community, means that we could determine the velocity of our reference frame by the laws of physics. This is contrary to the Principle of Galilean Relativity and the Principle of Einstein's Relativity. The Galilean Relativity proposes that only relative motion matters, and there is no mechanical experiment that you could perform to determine the velocity of your reference frame. Einstein's Principle of Relativity expands on Galileo's Relativity by proposing not only does relative motion matter but he also proposes that there is no mechanical, optical or electromagnetic experiment that you could perform that could determine the velocity of your reference frame. Einstein stated that you cannot measure your velocity with respect to nothing.

This experiment clearly shows that you can use optical and electromagnetic principles to determine the velocity of your reference frame.

This experiment also teaches that not only is there an absolute reference frame—the aether but there is no absolute inertial frame—a reference frame that is traveling in a straight line with a constant velocity.

I arrived at this conclusion as I was conducting this experiment. There is no place in the something universe where an inertial frame could exist, because there will always be some force acting on the inertial frame which will change its velocity and its direction of travel. The reason for this is because the center of mass of all the matter in the something universe will always exert a force on any object that moves in the something universe. If the inertial frame tries to move into the nothing universe, it would only be extending the bounds of the something universe; however, the center of mass of all the matter in the something universe will still exert an attractive force on the inertial frame.

It is ironic that before I conducted this experiment, I never gave it a thought, but I believed that there was an absolute inertial frame. After this experiment I believe that there is no absolute inertial frame, but there is an absolute reference frame—the aether.

# 15.
# Mass and Energy

We have seen from the two state universe theory that mass and energy are the basic building blocks of the something universe. Mass has two basic building blocks associated with it. These two basic building blocks of mass are complementary consisting of the positive particle and the negative particle. Energy also has two basic building blocks associated with it. These building blocks are also complementary and consist of the north pole and the south pole complements.

Mass and energy are so closely associated with each other that we must be careful to distinguish between their basic forms, such as basic mass and basic energy and their combined forms, such as complex mass and complex energy; otherwise, energy could appear to acquire mass properties and mass could appear to acquire energy properties. Mass and energy are different; they don't have the same properties.

Complex mass, such as the earth and the moon, consists of mass and energy so closely intertwined that this form of mass is erroneously used interchangeably with basic mass. Mass, which is so closely connected to energy, can't be considered to be basic or pure mass, because basic mass consists of only pure mass particles, and it doesn't contain any energy. Therefore, we must be careful to distinguish between these two forms of mass otherwise we can easily find ourselves involved in false reasoning about these two forms of mass, resulting in riddles, paradoxes and wrong conclusions about them. This can occur because of our false logic of giving energy the properties of mass.

Complex energy, such as kinetic energy and potential energy, consists of energy and mass so closely associated with each other that it, too, must be distinguished from basic energy. Basic or pure energy consists of the magnetic complements of the north pole and the south pole. This magnetic energy is pure because it doesn't contain any mass. Therefore, if we view complex energy to be the same as or identical to basic energy, we will also find ourselves involved with riddles, paradoxes and wrong conclusions about energy resulting from our false logic of giving mass the properties of energy.

Because mass and energy are always found closely associated with each other, it appears that they must be strongly attracted to each other in the same manner that complements are attracted to each other. Also, since mass and energy are the basic building blocks of the something universe, and since everything that exists in the universe must have a complement, we conclude that the complement of mass must be energy, because it is the only other building block of the something universe.

The above reasoning that mass and energy are complements is also verified from our findings that light consists of a wave particle space separated complement; and since a wave is energy and a particle is mass and the wave and the particle are complementary, then logically we must conclude that mass and energy are complements. Therefore, mass and energy can't exist in the same space at the same time. They can exist only if they are kept separated in time or space. Also, they must have complementary properties, not the same properties.

If mass and energy are complements that can exist only if they are kept separated in space or time, how can we resolve the findings that mass appears to increase with velocity? If mass in motion has energy, then it is self evident that both the mass and the energy must be in existence at the same time. And if mass and energy are complements and they both exist at the same time, we can only conclude that it is the energy that increases with velocity, not the mass. The mass remains constant but space separated from the energy. See conclusions 26, 27 and 28 in the appendix.

The three and four dimensional time-space universe theories do not distinguish between basic mass and energy and complex mass and complex energy. The four dimensional time-space universe theory in particular considers complex mass to be mass, and it considers mass to be interchangeable with energy. It even proposes an equation that shows exactly how they are interchanged. The equation proposed is the famous $E = mc^2$ formula. Where E equals energy, m equals mass, which is not defined, and c is the velocity of light. The four dimensional time-space universe theory doesn't take the complementary nature of mass and energy into consideration. Therefore, when basic mass moves with a velocity and it acquires energy, they give the acquired energy mass properties and conclude that the mass increased with velocity.

We now know that mass and energy are complements, and the basic mass doesn't increase with velocity, it merely becomes closely associated with the acquired energy. This acquired energy is the complement of the basic mass, and since they both exist at the same time they must be kept separated in space and they must be distinguished from each other.

Based on the four dimensional time-space universe theory, scientists believe that the mass does increase with the velocity, and they describe this increase in mass by an exact mathematical equation which shows exactly how much it increases with the velocity. They express the increase in mass to be equal to the rest mass divided by the square root of one minus the velocity squared divided by the speed of light squared.

College textbooks often describe a mass spectrograph type of experiment that they claim verifies the conclusion that mass increases with velocity. The experiment explains that if a charged particle, such as an electron is injected into a magnetic field, the particle will receive forces on it that causes the particle to make a circular path in the magnetic field, as shown in Figure 10.

Figure 10.

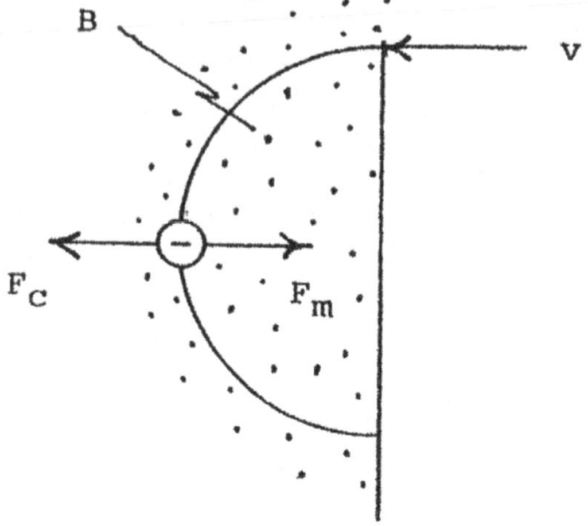


When an electron is injected into a magnetic field B. The forces acting on the electron are the centrifugal force $F_c$ and the magnetic force $F_m$. These forces are equal and opposite, and they cause the electron to travel in a circular path. If these forces are equated and the radius of the circle calculated, we find that the radius of the circle is proportional to the mass and velocity. If it is assumed that the mass is constant, then the radius will be directly proportional to the velocity. At low velocities this equation for the radius of the circle appears to be true. However, if the experiment is performed using high velocities approaching the speed of light, they find that the radius doesn't increase in a linear manner with velocity. That is, if the particle is injected into the device at some high velocity, and then another particle is injected into the device at double the speed, the radius of the particle should double. However, the radius doesn't double, it increases much more than it should. The scientists reason that since the radius is proportional to the mass and the velocity, then the increased radius results because the mass is increasing with the velocity.

When the experiment is conducted at velocities approaching the speed of light, and the value for the mass predicted from the relativity equation is inserted in the radius equation, then the radius is exactly as they calculate it should be. Based on these findings, the scientists claim that these experimental results clearly show that mass increases with velocity.

The equation for the radius used by the scientists, however, fails to take into consideration two very basic scientific principles. One of the principles was proposed by Maxwell from the findings of his famous equations. He states that an accelerating charged particle should radiate energy. The charged particle used in the experiment is accelerating because the direction of the velocity is constantly changing as it travels in a circular orbit. Therefore, the charged particle should radiate energy. If the particle radiates energy then the loss in energy of the particle will affect the radius of the circular path. However, the equation used in the experiment ignores this principle, because nowhere is this lost energy of radiation taken into consideration in the equation for the increase of mass with velocity.

The second very basic principle not taken into consideration is a principle, described in electromagnetic theory, that a charged particle in motion is a current, and a current always produces a magnetic field. In the experiment, the revolving charged particle is a current, which produces a magnetic field that is in a direction opposite to the direction of the applied magnet field. This opposing magnetic field results in a reduction of the applied magnetic field that the charged particle revolves in. See Figure 11.

In Figure 11, an electron having a velocity v is injected into the applied magnetic field B, represented as dots and directed out of the page. Thusly directed, it will cause the electron to move in a counter clockwise circular path in the applied magnetic field B. This counter clockwise rotation of the electron is the same as a current i revolving in a clockwise direction. The reason that a current flows in the opposite direction to the flow of electrons dates back to the early days of electricity.

Figure 11.

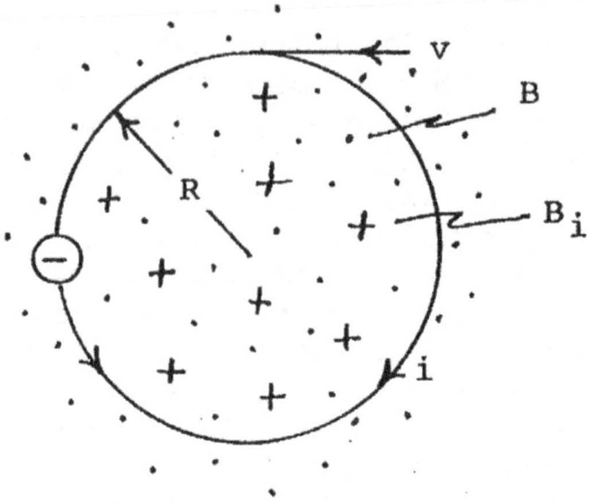


At that time the scientists didn't know what electricity was, so they assumed that current flowed from the positive side of a battery to the negative side of the battery. It wasn't until the vacuum tube was invented, many years later, that they realized that it was the flow of electrons in the vacuum tube that causes the tube to operate. However, since many textbooks were written using the convention that a current flows from the positive terminal to the negative terminal, which is opposite to the flow of electrons in a vacuum tube circuit; a majority of the scientists decided that it would be too confusing to change the textbooks. So they decided to keep the old system.

They are able to do this because they assume that current is the flow of positive particles. Therefore, the scientist uses the convention that a current flows in the opposite direction to the flow of electrons. This convention of the electron flowing in a counterclockwise direction and the current i flowing in the clockwise direction is shown in Figure 11. With the current flowing in a clockwise direction and using the right hand rule of electromagnetic theory, we find that the current i will generate an induced magnetic field $B_i$ as shown. $B_i$ is represented as crosses, and it is directed into the paper. These two magnetic fields buck each other resulting in the applied magnetic field B becoming weaker.

A weaker applied field, interacting with the moving charged particle, causes the radius of the circular path of the electron to increase. And an increased radius would indicate, from the four dimensional time-space universe theory, that the mass of the charged particle increased. However, this is not true since the increased radius results because the magnetic field became weaker, not because the mass increased. There can be no question that the moving charged particle generates a magnetic field. This effect is well known since it is a very basic principle of electromagnetic theory, but this effect is not taken into consideration in the experiment described in the textbooks.

Some knowledgeable scientists will try to pooh pooh this well-known principle of the applied magnetic field becoming weaker because of the bucking effect of the induced magnetic field generated by the moving charged particle. They claim that a single charged particle can't produce a magnetic field that can affect its motion while traveling in a magnetic field. However, if you ask the question, suppose that instead of one charged particle being injected into the device, numerous charged particles are injected into the device, wouldn't the many charged particles traveling in the magnetic field constitute a current, and wouldn't this generated current produce a magnetic field that opposes the applied magnetic field that the charged particles are moving in, and with a weaker magnetic field wouldn't the numerous moving charged particles move in a circular path that has a larger radius? The scientists will have no answer for this question because they interpret the increase in radius of the charged particles to be caused by the increase in the mass of the moving charged particles. They completely ignore basic principles of physics. It is clear, however, from basic scientific principles that the mass is not increasing, rather the magnetic field is getting weaker due to the induced magnetic field produced by the revolving charged particle or particles. The scientists do not have an answer to the questions proposed here because they wrongfully give energy mass properties and call it mass.

In my book titled, the two state universe, I showed, mathematically, that if the two effects of the radiating energy and the induced magnetic field are taken into consideration an equation results that is identical to the equation derived from the four dimensional time-space universe theory, which shows that the mass increases with velocity. My identical equation, however, was derived by assuming that mass doesn't increase with velocity, and it doesn't violate any principles of physics.

If my identical equation is used in the experiment described in the textbooks to show that mass increases with velocity, the calculated radius from this equation would be in exact agreement with the experimental results described in the textbook experiment. It was not unexpected that the results would be in exact agreement with the experimental results because the two equations are identical.

The equation derived from the two state universe theory and the results it produced clearly shows that mass doesn't increase with velocity, and this equation was derived without violating any of the principles of physics.

The action of a revolving charged particle, such as an electron, in a mass spectrograph type device, described in the textbooks, is similar to the wave particle complement because a revolving electron and its associated magnetic field are complements. Because of this we can't assume that this complementary pair is only an electron that increases with velocity. If we do, we are led into riddles and paradoxes similar to that of the dual nature of light. The same can be said about the similar complement of mass and energy or the mass in motion complement. If we wrongfully conclude that this mass in motion complement is only a mass (by giving energy mass properties), we find ourselves with another seemingly unsolvable paradox suggesting that mass increases with velocity.

This is another example of how a paradox and a wrong conclusion can occur when viewing the universe from the three or four dimensional time-space universe theories. See conclusion 28 in the appendix.

In this chapter we concluded that mass and energy are complements, like the wave and the particle nature of light. Based on this conclusion, we were able to logically show that in the mass spectrograph device, mass does not increase with velocity; rather, the magnetic energy of the revolving particle increases with velocity. This magnetic energy accompanying the particle as a complement counteracts the applied magnetic field of the mass spectrograph causing an increase in the radius of the path taken by the charged particle.

We also found that a revolving charged particle in the mass spectrograph radiates energy, as it should based on Maxwell's equations. The loss of energy causes a decrease in the velocity of the charged particle which when combined with the induced magnetic field effect, described above, results in an equation identical to that derived from relativity, which gives the appearance of an increase in mass with velocity in the mass spectrograph. We now know, however, that the equation derived from the two state universe theory clearly shows that mass does not increase with velocity.

# 16.
# The Mass and Energy Complement

In the last chapter we discovered that mass and energy are complements. We also showed that mass doesn't increase with velocity. In this chapter we will examine the mass energy complement in more detail because it is the most important complement in the something universe.

Everything in the something universe consists of mass and energy. This complement also contains both the electrostatic complement and the electromagnetic complement. The electrostatic complement consists of the positive and negative particles, and the electromagnetic complement consists of the north pole and the south pole of a magnet.

Charles Augustin Coulomb (1736-1806) a French physicist was able to show, mathematically and experimentally, that the electric force between charged particles is directly proportional to the product of the magnitude of the charged particles and inversely proportional to the square of the distance between them. The particles repel each other if they are the same and they attract each other if the charges are different.

Coulomb also did work on magnetic poles, and derived a law for magnetism, which states that the force between two magnetic poles is directly proportional to the product of the strength of the poles and inversely proportional to the square of the distance between them. The force is one of attraction or repulsion. The poles attract if they are different and they repel if the poles are the same.

Later an equation was derived which unified the electrostatic charge Q and the electromagnetic field B. The equation states that the force F acting on a charged particle Q traveling with a velocity v in a magnetic field B is equal to Q times v times B. The direction of the force is perpendicular to the direction of the velocity and it is also perpendicular to the magnetic field. The direction of the force depends on whether the charged particle is positive or negative.

These forces are a very important part of the something universe. Because these forces and other forces, such as the centrifugal force, are the forces that keeps the complements, that make up the something universe, in existence.

All of the elements known to exist in the something universe are made up of various combinations of positive and negative particles. These particles can only exist in the basic unit of each element, the atom, if they are kept separated. It requires energy to keep these charged particles separated, because these positive and negative particles are complements, which have a natural attraction for each other.

The different atoms, which are the basic unit of each of the known elements, consist of a nucleus, which contains the positive particles, and electrons, which revolve around the nucleus. The number of positive particles contained in the nucleus determines what the element is. For example, a nucleus with one proton with one electron revolving around it is the hydrogen atom. The element uranium has 92 protons in the nucleus and 92 electrons revolving around the nucleus. Because the atom is neutral, it contains an equal number of positive and negative particles.

In addition to all of the positive particles, the nucleus also contains neutral particles, called neutrons, which are needed to prevent the nucleus from flying apart.

We have seen from Coulomb's law that like charges repel and unlike charges attract. If the nucleus contained only the positive particles, the positive particles would repel each other and cause the nucleus to fly apart. The neutral particles together with the positive particles create enough energy in the nucleus to prevent this from occurring. The energy contained in the nucleus, which prevents the particles from flying apart, is called the binding energy. Each of the elements requires a different binding energy to keep their nucleus contained. And the nucleus of each of the elements requires different numbers of the neutral particles to accomplish this.

The electrons are required to rotate around the nucleus with enough kinetic energy to prevent the electron from being attracted and pulled into the nucleus. There is a balance between the complementary force of attraction and the centrifugal force, caused by the revolving electrons, which allows the complementary particles and the atom to exist.

Maxwell's equations showed that an accelerating electron will radiate energy. How then does the accelerating electron, as it revolves around the nucleus, continue to revolve around the nucleus without radiating energy? If the electron radiates energy, as it should according to Maxwell's equations, it would lose energy and quickly be pulled into the nucleus.

Mainstream physics states that there are certain allowed orbits that the electron can rotate in without radiating energy. This sounds good, but there is no logical explanation given for why the electron can revolve in only these certain orbits without radiating energy.

The two state universe theory shows that if a revolving electron is prevented from changing its orbit, it will not radiate energy. This occurs because, an electron, as it revolves around the nucleus of the atom, will generates an electromagnetic field in such a direction that if the electron starts to radiate energy it will began to be pulled into the nucleus. However, when the electron moves towards the nucleus, the induced magnetic field generates a force on the electron that causes the electron to increase its velocity. An increased velocity increases the centrifugal force causing the electron to move away from the nucleus. This reaction in the atom, when it is in a steady state condition, without any external forces acting on the revolving electron, is what prevents the accelerating electron from radiating energy. Therefore, we see that an accelerating charged particle does not always radiate energy.

If the electron receives a burst of energy from some source, it can move to a higher orbit because it would have more energy. In the higher orbit the electron is not stable, and it can lose energy to radiation. When it does, the electron returns to its original stable orbit, and in so doing, gives up, by radiation, the energy it acquired when it moved to the higher orbit. The radiated energy will be given off at a particular frequency, depending on the rotation of the electron. If the frequency given off is in the frequency range of visible light, then the electron will give off visible light as it returns to its original orbit.

The two state universe theory also prevents the electron from occupying the space that contains the magnetic energy because the electron mass and the energy are complements and they cannot occupy the same space. This is probably why the atom contains certain allowed orbits. These would be the orbits the electron can revolve around in because there is no magnetic energy present in these orbits, and the electron orbit cannot change because the forces acting on the electron are in balance. These are the conditions needed to prevent the accelerating electron from changing orbits and radiating energy.

We explained above that the nucleus contains binding energy, which prevents the nucleus from flying apart. We also now know that mass is not energy, it is the complement of energy, and it doesn't change into energy, nor does it increase with energy. How then does the atom bomb operate to give off energy?

In the atom bomb slow moving neutrons split the uranium atom, which causes the unstable nucleus to split into two or more lighter elements. The total binding energy of the newly formed elements have less energy than was contained in the original nucleus of the uranium atom that was split. It is the difference in the binding energy of the original uranium nucleus and the lesser total binding energy of the two or more lighter elements that causes the atomic bomb explosion.

Therefore, we see that mass not only doesn't increase with velocity it also doesn't change into energy in the atom bomb. After the explosion of the atom bomb, there is the same number of particles after the explosion as there were before the explosion, and the basic mass particles neither increases nor decreases in mass. Also the total energy contained in the atom bomb before the explosion is the same as the energy given off in the explosion plus the binding energy of the remaining particles given off after the explosion.

Another conclusion we can deduce from our findings in the last chapter is that as a charged particle revolves faster and faster in a magnetic field, the radius of curvature of its circular path increases with the velocity. When the charged particle reaches the speed of light the radius increases to infinity. This means that the charged particle is traveling in a straight line unaffected by the magnetic field. Thus electromagnetic radiation is a complement of a charged particle and its generated magnetic field traveling at least at the speed of light, and this electromagnetic complement as it travels through the aether is not affected by a magnetic field.

Also, because mass doesn't increase with velocity, there is no reason found in the two state universe theory that prevents a mass from traveling faster than the speed of light. If the aether space or the points in space are in the nothing state, there is nothing to offer resistance to the moving particle. With enough energy given to the particle there should be no reason why the particle can't travel faster than the speed of light. This is especially true for neutral particles because they would not have a magnetic field surrounding them, which could possibly offer resistance to the moving particle. We will examine these findings in more detail in the next chapter.

In this chapter we found that the mass and the energy complements are the building blocks of the something universe. Everything that we could feel, see and perceive as being real is composed of the mass and energy complement. We saw that this complement includes the electrostatic and the electromagnetic complements. We also described how the electrostatic and the electromagnetic complements were unified in a single equation. However, gravity doesn't appear to fit in with these two forces, and it appears to operate in a different manner. We will examine gravity in more detail later to find out where and how it fits in with the mass and energy complement.

# 17.
# Can a Particle Travel Faster Than Light

In the last chapter we found that it is possible for a particle to travel faster than the speed of light, because the mass of the particle doesn't increase with velocity. Also, when we solved the equation for the time of the two paths in the Michelson Morley experiment, we found that one of the solutions showed that the time it takes the light to travel along one of the paths, through the aether, was equal to zero.

This solution means that the light would have to be traveling at an infinite speed, and if the light or a wave with its associated particle travels with an infinite speed through the aether, then there would be no interference in the interferometer.

This solution derived from the two state universe theory suggests that light or a particle with its associated wave could possibly travel through the aether at an infinite speed. However, we know that the velocity of light through the aether is dependent upon the characteristics of the aether, but we don't know if a particle can travel faster than the speed of light through the aether. And if the particle can travel faster than the speed of light, we don't know if the energy in the form of a wave, being highly attracted to the particle, could be carried through the aether by the fast moving particle.

In this chapter we will try to determine if a particle can travel faster than light, and if this fast moving particle can carry its associated energy along with it. There is a considerable amount of evidence available indicating that a particle can and does travel faster than the speed of light. Let us examine this evidence.

Because the mass of a particle is an invariant, it doesn't need to have an infinite force applied to it to travel faster than the speed of light.

Maxwell's equations show that an electromagnetic wave travels through the aether as a transverse wave with a speed of light. The two state universe theory shows that this speed is a constant that is determined by the characteristics of the aether. Maxwell's equations also show that a changing magnetic field produces a changing electric field and the changing electric field produces a changing magnetic field. This process continues as the electromagnetic wave travels through the aether. Maxwell believed that this is the way an electromagnetic wave, which is energy, travels through the aether. However, his equations show that only an electromagnetic wave travels through space in this manner. It doesn't explain how other forms of energy, such as the kinetic energy of a fast moving particle, travels through the aether. It is possible that the kinetic energy of the fast moving particle could be so strongly attracted to the particle, because of their complementary nature, that the energy would be carried along, through the aether, by the fast moving particle.

Therefore, we see that energy could travel through the aether in two possible ways. Let us first look at the way an electromagnetic wave travels through the aether, and try to determine if a particle could be associated with this wave in such a way that it could travel faster than the speed of light.

Maxwell showed that a changing magnetic field creates a changing electric field. This finding is logical and there is no reason to believe that this doesn't occur. The changing magnetic field is pure energy, which is not composed of any known particles, because pure energy is the complement of mass and, therefore, it does not contain any particles, but the changing magnetic field, which is pure energy, produces a changing electric field.

If a particle were carried along with the electromagnetic wave, it would have to be carried along by being associated with the electric field. In fact it is illogical to believe that an electric field can exist without charged particles being associated with it. An electric field can only be created by charged particles. And moving charged particles is a current, and we know that a current produces a magnetic field, which is energy. Therefore, it is reasonable to believe that the changing electric field described by Maxwell is created by charged particles.

If this is true, what are these charged particles? We know that these charged particles can't be electrons if the electromagnetic wave is light, because if an electron traveled at the speed of light, it would have a wavelength in the x-ray region of the electromagnetic spectrum. It would not have a wavelength in the visible region of the spectrum.

If the particle is not an electron what would this particle be, and where does it come from? In the two state universe theory we know that particles can be created in the G factor from nothing. However, when particles are created, they must be created as complements. If Maxwell's changing magnetic field produces a changing electric field which consists of charged particles, then the particles creating the electric field must have been created from the aether in the form of at least one complementary pair of one positive and one negative particle.

If the changing magnetic field initiates this creation of a positive and negative particle, then the particles would have to travel in opposite directions in the circular orbit constituting the electric field. (If the two particles travel in the same direction, their effects would cancel.) When the two particles meet at a point that is half way around the electric field orbit, they would come together and become annihilated giving off pure energy, which is magnetic energy.

This process repeats itself creating a changing magnetic field and then a changing electric field, as described in Maxwell's equations. This is probably how a point in space is changed to an energy state as the energy travels from point to point in the aether space. From this example it is seen that the points in space that are changed from the nothing state to the energy state when they receive these bursts of energy are not next to each other. They are separated by a distance, which depends upon the wavelength of the electromagnetic wave traveling in the aether. The higher the frequency of the electromagnetic wave the closer the points in space becomes that changes from the nothing state to the energy state.

Because the points in space that give off the energy created by the annihilated complementary created particles are dependent upon the frequency of the electromagnetic wave, it is reasonable to conclude that these bursts of energy are the bundles of energy or quanta described by Planck. These bursts of energy are the bundles of energy that travels through the aether as a result of the electromagnetic wave as it travels through the aether. These quanta of energy or photons would appear to be massless because the particles creating them are annihilated as the bundles of energy are created. We see from this example that the quanta of energy occurs twice during each cycle of the electromagnetic wave.

This process of a changing magnetic field creating a changing electric field consisting of complementary pairs of charged particles is repeated when the changing electric field creates a changing magnetic field. The only question remaining is, does the particles created in the electric field travel faster than the speed of light? The answer to this question is yes, because the particles travel along the envelope of the electric field, and because the particles travel a greater distance than the wave travels during the same time period, the velocity of the particles would be greater than the velocity of the wave.

If we calculate the velocity of the particles and the velocity of the wave, we find that the wave travels at the speed of light and the particles travels with a velocity that is 1.57 times the speed of light. Therefore, there is evidence that in an electromagnetic wave, the particles that create the electric field do travel faster than the speed of light.

Before we examine the second way energy could travel through the aether space, I would like to describe a number of interesting findings I discovered while examining the speed of particles in an electromagnetic wave.

If we examine the characteristics of the points in space, we find that the time it takes a point in space to change states is equal to the Planck time, which is 5.4 x $10^{-44}$ seconds. The distance between the points in space is equal to the Planck length, which is equal to 1.616 x $10^{-35}$ meters. If we divide the Planck length by the Planck time we get a velocity equal to the speed of light. Using the values described gives a speed of 2.993 x $10^{8}$ meters per second.

This finding is interesting because it suggests that there is a limiting frequency that an electromagnetic wave can have as it travels in the aether. This frequency would be equal to 1 divided by the Planck time. This limiting electromagnetic wave frequency is equal to 1.85 x $10^{43}$ hertz. And since we have a limiting frequency, we would also have a maximum photon energy that can travel through the aether space. This maximum photon energy would be equal to Planck's constant h multiplied times the frequency f. If we multiply Planck's constant times the maximum or limiting frequency that can travel as an electromagnetic wave through the aether space, we get a maximum photon energy of 12.21 x $10^{9}$ joules or 7.64 x $10^{28}$ electron volts. This is a huge amount of energy. However, this amount of energy is the maximum amount of energy that can be transmitted through the aether space in the form of a quantum of energy in an electromagnetic wave.

From the two state universe theory, we know that if energy can travel through the universe in the form of bundles of energy in a discontinuous manner, then the complement of energy traveling through the aether in a discontinuous manner would be energy traveling through the aether in a continuous manner.

Energy traveling through the aether in a continuous manner is the second way that energy can travel through the aether. How does energy travel through the aether in a continuous manner? Let us examine this form of energy travel.

In the electromagnetic wave, the changing electric field causes energy to travel through the aether in the form of bundles of energy in a discontinuous manner; however, the changing magnetic field is pure energy that travels through the aether as a wave. This energy wave travels through the aether continuously with the bundles of energy superimposed on it.

Also, kinetic energy in the form of a fast moving particle is another way that energy could travel through the aether in a continuous manner, and it is possible that the fast moving particle could travel faster than the speed of light. This could occur because the energy associated with a fast moving particle would be so highly attracted to the particle, due to their complementary nature, that a particle traveling faster than the speed of light could carry the highly attracted energy through the aether without the points in space changing states from the nothing state to the energy state; this could occur because the points in space would not be able to change states, from the nothing state to the energy state, fast enough to impede the motion of the kinetic energy of the fast moving particle.

If a particle were traveling at a speed greater than the speed of light, then the time it would take a point in space to change states would be too slow to affect the motion of the high-speed particle and its associated energy. The fast moving particle and its energy would, in effect, then be traveling through the aether space with the aether space remaining in the nothing state. In this state the aether could not offer any resistance to the fast moving particle and its energy nor affect its motion in the aether. And the energy traveling through the aether in this manner would not have to wave anything to travel through it.

There is also a considerable amount of evidence showing that a particle traveling through the aether in this manner could travel faster than the speed of light. Let us examine this evidence.

In 1924, Louis de Broglie proposed that particles of matter, such as protons, neutrons, electrons or other particles are wavelike as well as corpuscular in nature. He derived an equation for the wavelength of these particles, which states that the wavelength of a moving particle is equal to Planck's constant h divided by the mass m of the particle, and the velocity v of the moving particle. Since we now know that mass doesn't change with velocity, it is an invariant, we don't have to be concerned about relative mass, and because of this, we are not required to limit the velocity of a particle to the speed of light.

We know that cosmic rays have energies of at least $10^{19}$ electron volts (ev), which is equal to 1.6 joules. Cosmic rays are extremely high-speed particles from outer space that enter the earth's atmosphere from all directions. These particles are electrons and the nuclei of atoms.

If we use de Broglie's wavelength equation and assume the particle is an electron, we find that these cosmic ray particles would be traveling with a speed that is approximately $4.3 \times 10^6$ times the speed of light. This is an extremely high speed. If we assume the particle is a proton, the speed of the proton would be approximately $10^5$ times the speed of light. There are probably many other particles, such as the neutrino, that travels faster than the speed of light and possibly faster than the speed of the cosmic ray particles.

Millions of neutrinos pass through each square foot of the earth's surface every second. They are traveling at such high velocities that they pass right through the earth as if it did not exist. Neutrinos are created on our planet when high-speed cosmic rays collide with the earth's atmosphere. For years scientists believed that neutrinos didn't have any mass. However, they now believe that they do contain an extremely small amount of mass. They suggest that they have a mass that is only about one ten-millionth, or less, than the mass of an electron.

If the neutrinos, do indeed, have this small amount of mass and they are created by cosmic rays, which we saw have extremely high speeds—much higher than the speed of light, then if the neutrinos, with a much smaller mass than the cosmic ray particles, acquire even a small amount of the cosmic ray energy, they could have velocities much higher than the cosmic ray particles that created them.

Let us examine other possible evidence that suggests that particles can travel faster than the speed of light. We know from de Broglie's wavelength equation that an electron traveling at the speed of light could not create visible light, because an electron traveling at the speed of light would have a wavelength of approximately 3.4 x $10^{-12}$ meters. This wavelength would be much too small to give off visible light, because visible light has wavelengths of approximately 4 x $10^{-7}$ meters, for the violet region of light, to 7 x $10^{-7}$ meters, for the red region of light. We would need a particle with a much smaller mass than the electron to produce wavelengths in the visible spectrum.

If we calculate the mass of a particle that travels at the speed of light with a wavelength of 7 x $10^{-7}$ meters in the red region of visible light, we find that the mass of this particle would be equal to 3.1 x $10^{-36}$ kg. Because the electron has a mass of 9.1 x $10^{-31}$ kg, this particle would be 2.9 x $10^{5}$ times smaller than the electron. This particle could be a neutrino, which the scientists believe has a mass of one ten millionth the mass of the electron.

If we calculate the velocity needed by a neutrino to create light in the red region of the visible spectrum, we find, from the de Broglie wavelength equation using the wavelength of 7 x $10^{-7}$ meters, that the velocity of the neutrino would be 100 x $10^{8}$ m/sec. This velocity would be 33 times faster than the speed of light.

Since the electromagnetic spectrum consists of infinite numbers of different frequencies, it would not be logical to assume that each frequency of the electromagnetic spectrum is created by a different particle. It is more logical to assume that the infinite numbers of frequencies are created by only a small number of different particles, with each of the particles creating a different band of frequencies as the particle travels at different velocities.

Using this logic, if we now calculate the velocity needed by the same neutrino to create light in the violet region of the visible spectrum, we find from the de Broglie wavelength equation, using the wavelength of 4 x $10^{-7}$ meters, that the velocity of the neutrino would be 181 x $10^{8}$ m/sec. This velocity would be 60 times faster than the speed of light.

And if we assume that the neutrino travels at a velocity which creates a frequency with a wavelength of $10^{-12}$

meters in the x-ray region of the electromagnetic spectrum, we find that the neutrino would be traveling with a speed that is $24.3 \times 10^6$ times faster than the speed of light. This is over 24 million times faster than the speed of light, which is a very high velocity. The reason I assumed the frequency with a wavelength of $10^{-12}$ meters in this example is because beyond this frequency an electron traveling at the speed of light would also create a frequency in this wavelength range. The electron could create frequencies up to the frequencies in the cosmic ray region. We already calculated the speed an electron could have, as it traveled in the cosmic ray region of the electromagnetic spectrum. We found this speed would be $4.3 \times 10^6$ times the speed of light. If we calculate the wavelength of this cosmic ray, we find that it is equal to $5.6 \times 10^{-19}$ meters. And if the neutrino traveled fast enough to have this same wavelength, it would be traveling with a velocity that is $4.3 \times 10^{13}$ times faster than the speed of light.

In this chapter we found that energy can travel through the aether in two different ways. One of the ways it travels through the aether is in the form of an electromagnetic wave as described by Maxwell. In the electromagnetic wave we found that energy travels continuously in the magnetic part of the wave; however, energy also travels in a discontinuous manner in the form of bundles of energy, or quanta, that was generated by the electric part of the electromagnetic wave. The electric wave also contains at least one complementary pair of particles which we found travels faster than the speed of light.

The second way that energy can travel through the aether is in the form of kinetic energy, consisting of a fast moving particle and its associated energy. And since the particle and the energy are complements, they are strongly attracted to each other. When the particle travels faster than the speed of light it carries its associated energy, in a continuous manner, along with it. And because the particle carries the energy along with it, traveling faster than the speed of light, the aether space, because it can't respond fast enough, remains in the nothing state. In so doing, it allows the particle and the energy to travel faster than the speed of light without offering any resistance to the complementary pair of the mass and its associated energy.

Thus we see that a particle can and does travel faster than the speed of light; and there is no reason, that we know of from the two state universe theory, why a particle can't travel at any speed.

# 18
# Gravity

Galileo, one of the giants of science, died in 1642. Isaac Newton, another giant of science, was born in Woolsthorpe, Lincolnshire, on Christmas Day in the same year. In 1661, Newton entered Trinity College, Cambridge and graduated from Cambridge in 1665 with a Bachelor of Arts degree.

The bubonic plague broke out in London in late 1664, and by the summer of 1665 it caused Cambridge University, which is located about 50 miles north of London, to close its doors.

Newton returned to his birthplace in Woolsthorpe to wait out the plague. Cambridge remained closed for about eighteen months. These eighteen months were the most productive months of Newton's life. It was during this time that he conceived and worked out most of his great works. In mathematics he developed the binomial theorem, differential calculus and integral calculus. He developed his laws of mechanics, the theory of colors and optics, including light, and the law of universal gravitation.

Newton conceived of these great works when he was between the ages of 23 and 25. However, he kept most of this work secret until he wrote his famous book called the Principia, which consisted of three books. His works on mechanics and gravity was published in 1687, and his work on optics was published in 1704.

Newton was known to have said that if he had seen further than any man, it's because he was standing on the shoulders of giants. When Newton discovered and developed his famous theory of universal gravitation, he was standing on the shoulders of at least three giants—Copernicus, Kepler and Galileo.

Copernicus showed, contrary to the teachings of Aristotle and Ptolemy, that the earth is not the center of the universe, and contrary to all of our senses and logic, it is not at rest in the universe. He showed that the earth and all the planets travel around the sun. He also calculated the radius of the planetary orbits around the sun.

Kepler showed that the orbits of the planets around the sun were elliptical with the sun located at one of the focus of the elliptical orbits. He also formulated the laws for the motions of the planets in their elliptical orbits around the sun.

Galileo, the father of experimental science, through his experiments, showed that two objects, regardless of their masses, fall to the earth with the same acceleration. He supposedly dropped a light ball and a heavy ball off the tower of Pisa; the two balls landed on the ground at the same time. If two objects, such as a hammer and a feather, are dropped at the same time, in the absence of air resistance or other forces, the hammer and the feather will reach the ground at the same time. An astronaut on the moon conducted this experiment, and, indeed, the hammer and the feather did fall to the surface of the moon at the same time.

Galileo also discovered the principle of inertia experimentally showing that an object in motion will continue in motion in a straight line unless acted on by an external force.

Newton was certainly aware of all these findings when he left Cambridge to escape the plague. Newton discovered his law of universal gravitation during his stay at Woolsthorpe. He supposedly was sitting under an apple tree when an apple fell off the tree and hit him on the head. Newton was well aware that all objects fall to the ground everywhere on earth. He wondered how far above the earth this force extends.

He realized that the moon is being attracted to the earth by some force. He knew this because of Galileo's findings of inertia, which stated that an object in motion will remain in motion in a straight line unless acted on by an external force. Newton realized that if the moon was not attracted to the earth, the moon would fly off into space in a straight line. Instead it continues to fall towards the earth by the mutual attraction of the earth and the moon. He was also aware that the earth travels around the sun, so the earth and the sun must also be mutually attracted to each other. This reasoning led him to the idea of universal gravitation, which means that all objects are attracted to each other.

He wondered if the force that was attracting the apple to the earth was the same force that was attracting the moon to the earth. He reasoned that if this is true, then the force must get weaker the further away from the earth the object goes. He further reasoned that because the apple was small compared to the moon, the force of mutual attraction must be directly proportional to the product of the masses being attracted and inversely proportional to the square of the distance between them.

He knew that the acceleration of gravity on the surface of the earth was 32 feet per second per second. This was well known from the experiments done by Galileo. Galileo knew that all objects fall to the earth with the same acceleration. It doesn't matter how massive the objects are; they are all attracted to the earth with the same acceleration of 32 feet per second per second.

Newton was able to calculate the acceleration of gravity on the moon, caused by the attraction between the earth and the moon. This acceleration is equal to the velocity of the moon around the earth squared divided by the distance between the earth and the moon.

If the inverse square law was correct, then the ratio of the acceleration of gravity on the moon to the acceleration of gravity on the surface of the earth would have to be equal to the radius of the earth squared divided by the distance from the center of the earth to the moon squared.

In order to make this calculation the radius of the earth had to be known, and the distance to the moon had to be known. However, due to a mistake made by Newton about the size of the earth's radius, his calculations were not in agreement with the actual ratio of the gravity on the moon to the gravity on the surface of the earth. Because of the inaccuracy of the results, Newton supposedly set his idea of gravity aside for about twenty years. At that time he became aware of the true value of the radius of the earth. Upon learning the true value of the radius of the earth, he immediately returned to his work on gravity, and with this new information, he was able to get results that agreed exactly with the ratio of the acceleration of gravity on the moon to the acceleration of gravity on the surface of the earth.

These results confirmed his theory of gravity, and indeed the force that caused the apple to fall to the ground was the same force that keeps the moon in orbit around the earth.

This theory of gravity was a great step forward for science. However, there appears to be a number of problems with this theory. This theory of gravity doesn't appear to operate according to the other two forces of nature—the electrostatic force and the electromagnetic force.

The electrostatic force produces a force of attraction between oppositely charged objects, and the electromagnetic force produces a force of attraction between opposite poles of magnets.

Newton's theory of gravity is also in conflict with the two state universe theory, because the two state universe theory of complements shows that like parts of a real pair complement repel; unlike parts attract. See Law 5 in the appendix. Newton's theory of gravity shows that likes attract likes, that is, mass attracts mass. This finding is in conflict with Law 5 of the two state universe theory.

When Newton derived his theory of gravity, he failed to take into consideration the fact that the moon and the earth are complex masses consisting of basic mass and energy. The question then arises, what is attracting what? Does basic mass attract basic mass? Does energy attract energy? Or does mass attract energy?

We know from the two state universe theory that mass is the complement of energy. Every time we have a mass we find that it is associated with and attracted to energy. Therefore, in my book titled, the two state universe, I proposed that gravity is not the attraction between the masses or the attraction between the energy contained in the objects; rather, gravity is the attraction between mass and energy.

Therefore, from the two state universe theory, I proposed a new theory of gravity which states that the force of gravity is directly proportional to the mass times the energy and inversely proportional to the square of the distance between them. The mass referred to here is basic mass or rest mass not complex mass.

The inverse square law used in Newton's equation and the new two state universe gravitational equation is correct and it makes sense, because if we enclose an object containing basic mass and energy in an imaginary sphere, the energy that leaves the object and passes through the surface area of the imaginary sphere should pass through the surface area of the imaginary sphere pretty evenly distributed over the surface area of the sphere.

This occurs because energy traveling in the same direction away from an object repels itself causing the energy per unit area to become evenly distributed throughout the total surface area of the imaginary sphere. Even if the energy per unit area of the imaginary sphere is not quite evenly distributed over the surface area of the imaginary sphere, when it is located at the surface of the object, it would eventually become evenly distributed over the surface area of the imaginary sphere when we increase the size of the sphere, surrounding the object, by increasing its radius.

If we then continue to increase the radius of the imaginary sphere, the surface energy density of the sphere will become less and less as the radius becomes greater and greater, because less and less energy will pass through a unit surface area of the sphere, because the surface area of a sphere gets greater and greater in direct proportion to the square of the radius of the sphere. This causes the energy density to decrease inversely with the square of the radius or square of the distance from the center of the imaginary sphere enclosing the object. Therefore, energy density decreases inversely with the square of the distance from the center of an object consisting of basic mass and energy.

Newton also showed by using his newly discovered calculus that the total mass of an object could be considered to be located at the center of mass of the object. Thus, the inverse square law, which states that the force of gravity is inversely proportional to the square of the distance between the objects, is referring to the distance between the centers of mass of the objects; and the center of mass of a spherical object of uniform density is located at the center of the sphere.

Therefore, from the teachings of the two state universe theory it is clear that the force of gravity must be directly proportional to the mass times the energy and inversely proportional to the square of the distance between the centers of the objects.

How Newton's force of gravity actually operates is difficult to understand because his force of gravity depends upon an attractive force at a distance between two masses. We now know that the mass of the moon and the mass of the earth are not pure masses. Both the complex mass of the earth and the complex mass of the moon consist of basic mass plus energy. And from the two state universe theory, only the complements of the mass and the energy are attracted to each other. Newton's theory of gravity doesn't explain how two basic masses, separated by a distance, can be attracted to each other.

Also, the theory of gravity proposed by the four dimensional time-space universe theory does not offer a satisfactory explanation of gravity either, because it proposes that complex mass warps time and space. The problem with this proposal is that time is nothing—it has neither mass nor energy, and space is nothing—it too contains neither mass nor energy. They both are the basic units of the nothing universe.

It is clear from Einstein's special theory of relativity that he didn't believe in an aether space. He believed that everything is relative because space is nothing. Therefore, you can't reference anything to nothing, so everything must be relative to something else.

But, Einstein doesn't explain, in his general theory of relativity, how nothing, which doesn't have any characteristics, can be warped. When he presented his paper on the general theory of relativity, he used complicated mathematics, including tensor calculus, to explain it. At that time only a couple of scientists could understand the theory. But, he didn't explain in his mathematical concepts, what physical characteristics of time and space allows them to be warped.

Even if we consider space to be the aether space proposed by the scientists, this does not help the problem either. Because in the nothing state, the aether is nothing, so it can't be warped.

Further, Einstein's special theory of relativity is based on the idea that there is no aether space. The existence of an aether space invalidates Einstein's special theory of relativity, and it doesn't help the theory of gravity derived from his general theory of relativity.

# 19.
# Gravity in the Two State Universe

In the last chapter, we found that Newton failed to realize that there is a considerable amount of energy contained in the apple, the moon, the earth, the sun and other objects. He was certainly not aware of the atomic theory and the considerable amount of binding energy contained in these objects. Therefore, he considered these objects to consist of only mass.

He failed to realize that these objects contained both mass and energy, so he logically assumed that mass was attracted to mass. From the two state universe theory, we now know that mass is attracted to energy; so logically, the force of gravity must result because of the natural attraction between mass and energy.

We also found in the last chapter that the theory of gravity proposed from the four dimensional time-space universe theory does not offer a more satisfactory explanation of gravity either, because this theory of gravity proposes that a mass warps space-time which causes the effect of gravity. However, we pointed out that time and space does not have any physical characteristics that can be warped.

Only the something universe, which is the complement of the nothing universe, has components that can be warped. The main components of the something universe is the mass and energy complement, and it is the natural attraction between these two complements that results in the force of gravity.

Let us now examine the something part of the something nothing complement to determine exactly how and where gravity fits in the two state universe. See Figure 12.

Figure 12.

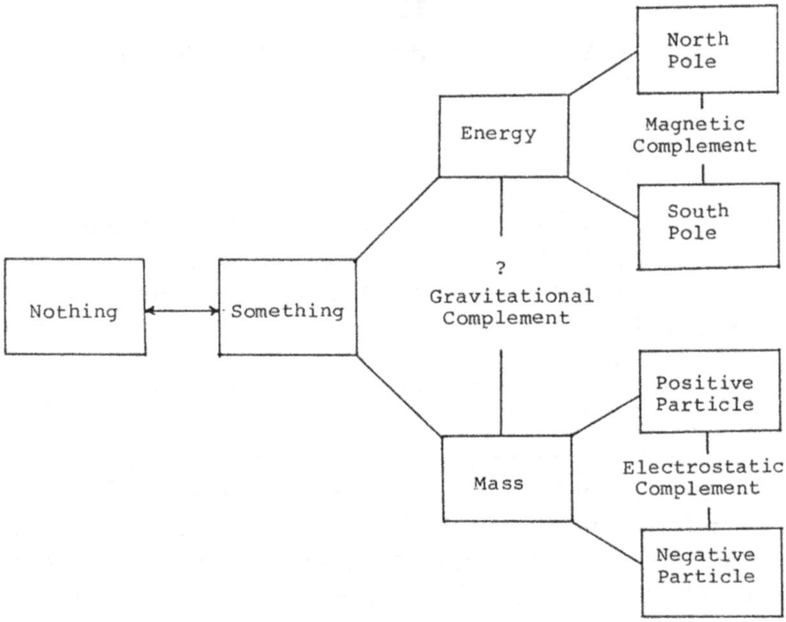


In Figure 12 we show a block diagram of the something nothing complement. However, in the figure we don't show the complements associated with the nothing part of the complement, because we know that gravity is not associated with the nothing part of the complement. Gravity has characteristics that cause the mass and energy to attract each other. So gravity must fit in the something part of the something nothing complement.

In the figure we show that the something part of the something nothing complement consists of the complements of mass and energy. Associated with energy is the magnetic complement consisting of the north and south pole complementary parts. Considered as basic energy, these two complementary poles are attracted to each other.

Associated with mass is the electrostatic complement consisting of the positive and negative complementary parts. Considered as basic mass, these two complementary particles are also attracted to each other.

Since mass and energy are complements they, too, must be attracted to each other. It is this attraction between the complements of mass and energy that accounts for the mysterious gravitational force. It is clear from the figure that this is the only place that gravity could fit in the something part of the something nothing complement, and operate in a manner that is in agreement with the way the other two complements (the electrostatic and the electromagnetic complements) operate in the two state universe.

As a result of this logic, it was proposed that the force of gravity is directly proportional to the mass times the energy, and it is inversely proportional to the square of the distance between these two complementary parts.

In order to use this new gravitational equation, it was necessary to find the value of the gravitational constant, which is needed to determine the basic mass and energy for all of the elements that exist in the something universe. Knowing these quantities we could compare the force of gravity proposed by Newton with that derived from the two state universe theory.

We found that the gravitational constant $K_g$ is equal to G/2 where $K_g$ is the new gravitational constant and G is equal to the gravitational constant found for Newton's gravitational equation.

To determine the basic mass contained in all the elements found naturally on earth, we assumed that the basic particles that make up the elements do not change in mass regardless of whether they are bounded in the atom or not—mass is constant. We assumed that all the elements are composed of multiples of the three basic mass particles—the proton, the neutron and the electron.

We also assumed that the basic particle is the deuterium atom, which is made up of a proton and a neutron which makes up the nucleus and an electron that revolves around the nucleus. We further assumed that the basic mass of the deuterium atom was equal to 2 atomic mass units. This assumption was necessary because the parameters used to derive the atomic mass of the proton, neutron and the electron was not known. Further laboratory work is needed to determine the exact rest mass of each of the basic mass particles.

Assuming that each of the elements are made up from these basic mass units, we solved for the average basic mass for each of the elements by multiplying the mass of the deuterium atom times the atomic number of the element. Since the deuterium atom has a mass of 2 atomic mass units, it is only necessary to multiply 2 times the atomic number of the element to determine its average mass in atomic mass units. We refer to the average mass because of the isotope effects. Atoms that have the same atomic number but their atomic weights are different are isotopes of the same element. The different atomic weights result because the nucleus of the same element contains different numbers of neutrons.

Deviations from the average value in nature occurs when the nucleus of the element, for whatever reason, gains or loses a neutron. Since the neutron has a mass of about 1 amu, we could determine the actual mass of the element by adding or subtracting 1 amu for each of the number of excess or deficient neutrons contained in the nucleus from the average mass determined for that element.

Knowing the basic mass for each of the elements, we were able to solve for the energy contained in each of the elements from an equation we derived by equating Newton's force of gravity with the force of gravity derived for the new theory of gravity.

In my book, the two state universe, I calculated all these values and tabulated them. Using these values, I then calculated the force of gravity between elements with the same atomic number and for elements with different atomic numbers. The results showed that the force of gravity between elements with the same atomic number was in exact agreement with the results derived from Newton's gravitational equation. However, when calculations were made between elements with different atomic numbers, we found that there was a close agreement with the results from

Newton's equation; however, the agreement was not exactly the same. We found a small error between the results from the two different equations.

Because we found this error between the two equations, we needed to find another way to determine the mass and energy for each of the elements. We realized that we could use the mass spectrograph to determine these values because in it, complex mass appears to increase with velocity, and it is proportional to the gravitational mass when measured at the same place on earth.

The mass spectrograph is used to determine the mass of the elements, particularly the mass of the isotopes of the elements. Although the atomic weights (given in atomic weight tables) were derived primarily by chemical means, they can be derived more precisely from a mass spectrograph.

In my book titled, the two state universe, I derived two equations from the mass spectrograph theory. One of the equations derives the atomic weight for each of the elements. The second equation derives the energy for each of the elements. We also derived two equations, which corrects for the isotope effect.

Using these equations we derived the atomic weights for each of the elements, and the energy contained in each of the elements. We tabulated the results and plotted graphs of the results. The results showed that the atomic weights calculated from the equations were in good agreement with the atomic weights given in published tables. And the atomic energy derived from the spectrograph calculations were in good agreement with the energy equation derived by equating Newton's equation with the two state universe theory of gravity.

Although the results were in good agreement, they were not in exact agreement; however, when the isotope corrections were taken into consideration in all of the equations, the results showed an amazing correlation between the atomic weights and the energies for each of the elements. Therefore, the spectrograph results were in good agreement with the results derived from the new theory of gravity.

These results were surprising and unexpected because they suggest that Newton's law of gravity may be in error by a small but real amount because the two equations, derived from the two state universe theory, using two different approaches, are in good agreement with each other; but they differ from Newton's equation by a small amount, suggesting that there could be a small error in Newton's equation, which varies depending on the combination of elements used. The truth to the above findings can only be determined through more careful laboratory work.

# 20.
# Black Holes

The idea about the existence of black holes is not new. It has been around for more than 200 years. Two scientists described the possibility of their existence independently of each other at about the same time.

One of the scientists was John Michell from Cambridge. He wrote a paper in 1783 in which he proposed that if a star is very dense and massive, it would have a gravitational attraction that is so strong that nothing, not even light could escape from this strong gravitational attraction. At that time light was considered to consist of particles. The particles of light would be so strongly attracted to the star that the gravitational attraction would prevent these particles of light from escaping its gravitational attraction. Therefore, the star would be black because it couldn't be seen because no light could escape from it.

A short time later the French scientist Laplace proposed essentially the same idea, but apparently, for whatever reason, he didn't pursue this idea. Could it be that he changed his mind about the existence of black holes?

Although the idea of black holes has been known for many years, it wasn't until 1969 that John A. Wheeler of Princeton University applied the name black hole to this idea.

When the wave theory of light became popular, it became difficult for the scientists to understand how gravity could affect the wave nature of light. It wasn't until Einstein proposed his general theory of relativity in 1915, that the scientists were able to understand how light, traveling as an electromagnetic wave, could be attracted by gravity.

Einstein proposed that gravity was the warping of space-time by a mass. Since the space-time was warped around the mass, any object, including light that approached the mass, would naturally follow the curved space-time leading to the mass. In this manner it would appear that the object, including the light, was being attracted by the mass, but instead, they were merely following the shortest path to the mass.

To understand how a star could eventually become a black hole we need to examine the life cycle of a star. A star is formed from a massive amount of gas consisting primarily of hydrogen. This large amount of gas becomes denser as its molecules are attracted to each other by gravity. As the volume of this gas gets smaller and smaller, the molecules collide with each other more often causing the temperature of the gas to increase. When the volume of gas gets hot enough, the hydrogen atoms began to fuse together to form helium. This reaction is the same reaction that occurs in a hydrogen bomb.

When this occurs, a considerable amount of energy is released causing the forming star to exist, giving off heat and light. This occurs because the massive amount of energy generated within the star creates a great pressure, which counteracts the force of gravity, resulting in the star becoming stable. The star could exist in this balanced state for many millions of years, with the life of the star depending on the amount of gas present in the star.

After the star uses up all of its nuclear fuel, it begins to cool off causing the star to contract. The contraction causes the star to become more dense and massive. However, because of a force caused by Paulie's exclusion principle, resulting from the force created by the repulsive force of electrons, the star could reach a balance and maintain a constant size, because of the balance between the force of gravity and the force resulting from the exclusion principle.

An Indian graduate student named Subrahmanyan Chandrasekhar, however, calculated that a star that is about one and a half times more massive than the size of the sun would not be able to prevent itself from collapsing against its own gravity because there is a limit to the amount of force that the exclusion principle could create. This mass of one and a half times the mass of the sun is known as the Chandrasekhar limit.

A Russian scientist named Lev Davidovich Landau, however, suggested that there was another state that the star could reach because of the repulsion between the protons and neutrons in the star. Stars in this state would be smaller and more massive. These stars are called neutron stars because they are made of neutrons.

If, however, the mass of the star exceeded the Chandrasekhar limit, the star would continue to collapse and become even more massive and compact than the neutron star. Just how far the star could collapse was not known at that time.

Robert Oppenheimer, the American Scientist who headed the atomic bomb project during the Second World War, proposed a solution to this problem. His work showed that the star would continue to collapse. And at some point during the collapse, the star would become so massive that the electromagnetic rays of light could not escape from the warped space-time surrounding the star. The size or the surface area of the star at that point of collapse is called the event horizon. The star could continue to collapse, but the event horizon would be the point where anything that approaches and enters the event horizon cannot escape from the strong gravitational effects of the collapsed star. Even the electromagnetic energy of light cannot escape falling inside the black hole, if it passed inside the event horizon.

Work done by Stephen Hawkins and Roger Penrose showed that the star could collapse to a point. They described this collapse to be a singularity, where there is an infinite curvature of space-time and an infinite density of the mass of the star inside the black hole. Such a star that collapsed to a singularity could not be seen because nothing could escape from the black hole.

How then could a black hole be detected? Scientists point out that, although the black hole can't be seen its effects can be observed. For example, as the black hole absorbs matter from a nearby star, this matter speeding towards the black hole gives off x-rays, which can be detected. Also, a star that is attracted to the black hole could be revolving around the black hole. The black hole can't be seen, but the star revolving around it can be observed, indicating the possibility of the presence of a black hole.

Scientists believe that they have observed a number of black holes. One is Cygnus x-1, another was found with the Hubble Space Telescope in the galaxy Centaurus A and in 1917, astronomers discovered a possible black hole in M 87, one of the largest and brightest galaxies located about 50 million light years from earth. A light year being the distance light travels in one year.

The black hole was discovered when the astronomers discovered a large cone shaped beam of light extending many thousands of light years out from the axis of M 87. The Hubble Space Telescope confirmed this possible black hole when it found a 500 light year wide disk of matter in orbit 60 light years away from the black hole. The disk was spinning at about 550 kilometers a second. The astronomers are at a loss to explain what possible mass could be large enough to prevent this spinning disk from flying apart. Recently astronomers also suspect that there is a possible black hole at the center of our Milky Way Galaxy.

The above evidence and logic gives us good reason to believe that black holes could exist. However, there are many problems with the above evidence and logic, which requires further study. Therefore, in the next two chapters we will try to answer the following questions about black holes using mainstream physics and the two state universe theory. The questions we need to examine in more detail are as follows:

Can a black hole exist? Can a black hole exist as described above by mainstream physics? Can black holes continue to draw in huge amounts of matter and energy and compress this matter and energy into a singularity of infinite density with an infinite curvature of space-time? And if indeed black holes do exist can they continue to exist indefinitely?

# 21.
# Can Black Holes Exist?

A scientific proverb says that a thousand experiments can't prove a theory right, but one experiment can prove a theory wrong. Presently there is no known scientifically sound experiment that proves the black hole theory wrong. Logically, however, it is difficult to understand, either from mainstream physics or from the two state universe theory, how a black hole can exist.

For example, a theory proposed by Dr. Stephen Hawking explains how a black hole can absorb negative energy causing the black hole to evaporate and vanish. This theory is creating doubt and confusion in the scientific community about black holes because any information contained in the black hole would be lost to the universe when the black hole vanishes. When information is lost to the universe it leads to a conflict with the principle of determinism of quantum mechanics.

The strong gravitational attraction of a black hole, however, can also absorb positive energy. If it absorbs enough positive energy to cause a reversal of the process that created it, the black hole would be prevented from vanishing, and the problem of the vanishing information wouldn't occur.

Although the black hole theory appears to be in a state of doubt and confusion, the scientific community favors it. And as we have seen from the last chapter, over the past 200 years, the black hole theory appeared to be in and out of favor depending upon the theory of physics popular at the time.

For example, when Newton's particle theory of light was in favor, scientists believed in the black hole theory because they could understand how gravity could attract a particle, but when Huygens wave theory of light was in favor, scientists couldn't understand how gravity could affect the wave property of light. And when Einstein's general theory of relativity was proposed in 1915, scientists again favored the black hole theory.

This was because Einstein proposed that gravity was not a force at a distance; rather, it was the warping of time-space by a mass. If time-space is warped, scientists could again understand how a wave could be affected by gravity, because the wave would have to follow the warped time-space as it travels through it.

And recently, with the discovery by the Hubble Space Telescope that a black hole may be in the M 87 galaxy, the scientific community strongly favored the black hole theory. But the evidence found is not conclusive, because scientific principles other than gravity can explain the strong force needed to keep the 1.2 million mph spinning disk of primarily ionized hydrogen and oxygen from flying apart. For example, the electrostatic and electromagnetic forces generated by the ionized hydrogen and oxygen gas could possibly be strong enough to prevent the revolving disk of ionized gas found in M 87 from flying apart.

Hawking proposed his black hole theory to explain the radiation given off by black holes. Scientists couldn't understand how a black hole could radiate energy. By definition the gravitational attraction at the event horizon of the black hole is so powerful that nothing not even light can escape from inside the black hole.

The event horizon, introduced earlier, is the boundary of the black hole where the escape velocity of anything at the boundary would have to travel with a velocity greater than the speed of light to escape falling into the black hole. According to Einstein, because nothing can travel faster than the speed of light, nothing could escape from the black hole not even light.

Hawking suggests that negative energy, in the form of one of a pair of virtual particles constantly being created in the vicinity of the event horizon of the black hole, could be absorbed by the black hole. The virtual particle pair consists of a particle and an antiparticle. According to quantum mechanics, and the two state universe theory, these particles can be created from nothing, because one particle has negative energy and the other particle has positive energy. Such particles exist for only a brief time, and then they recombine and are annihilated. Although they can't be detected, the particle and the antiparticle pair are called virtual particles because their effects on other particles can be measured.

Hawking believes that the negative energy particle or antiparticle created in the vicinity of the event horizon of the black hole could be absorbed by the black hole. The remaining positive energy particle or antiparticle would then become a real particle because its negative energy particle would not be available to combine with. Therefore, it would be free to escape from the black hole. The escape of the real particle would appear to be radiation coming from inside of the black hole. But this radiation is not coming from inside the black hole; rather, it is coming from outside. This apparent radiation is called Hawking radiation.

The absorbed negative energy combines with the mass of the black hole causing it to shrink. Eventually, the black hole evaporates and vanishes in a final burst of energy. This energy comes from the massive amount of energy created when the antiparticles, that were absorbed by the black hole, combines with the real particles contained in the black hole. According to Hawking, the black hole takes millions and sometimes billions of years by this process for it to absorb enough energy to evaporate and vanish.

The idea of a vanishing black hole is creating doubt in the scientific community about the black hole theory, because scientists can't understand how any information contained in the black hole could vanish from the universe. The idea of information vanishing from the universe is in conflict with the determinism principle of quantum mechanics. This states, in effect, that if all the particles and their present states are known in the present then theoretically their future states could be calculated and known. The scientific community is feverishly trying to understand the paradox of the lost information.

Hawking's theory describes what happens to a black hole when it absorbs only negative energy. However, it doesn't explain what happens to the black hole when it absorbs a massive amount of positive energy. A black hole not only can absorb negative energy, but it can also absorb positive energy. And if a black hole absorbs positive energy, theoretically it would be possible for the black hole to radiate internal energy and light when it had absorbed enough positive energy to create an internal pressure great enough to overcome the gravitational attraction of the black hole. Where would all this positive energy come from to create such a pressure?

In an experiment to determine the intensity of the radio waves emitted from our galaxy, in 1964, two radio astronomers, Arno A. Penzias and Robert W. Wilson, discovered a considerable amount of noise in the microwave band of 7.35 centimeters which was independent of direction. Moreover, the noise didn't vary with the time of day or the seasons, indicating that it was coming from a vast section of the universe. They determined that the equivalent temperature of this radio noise was between 2.5 and 4.5 degrees Kelvin.

Later, together with the work of other scientists, it was found that this microwave radiation was due to the energy that was left over from the big bang, and the temperature of the universe is about 3 degrees Kelvin. Although this temperature is quite low—only a few degrees above absolute zero—it represents a considerable amount of energy because this energy permeates the entire universe.

Because the event horizon of any black hole would be in direct contact with this energy, the black hole could easily absorb enough energy to replace the energy that was lost during its creation. At this point, the gravitational attraction of the black hole could no longer prevent energy or light from internally escaping. And anything in the vicinity of the event horizon would no longer have to travel faster than the speed of light to escape. In effect, the process that created the black hole is reversed, because the pressure created by the absorbed energy would counteract the strong gravitational attraction, allowing anything—including light and energy—to again escape.

If the black hole absorbs the energy, which surrounds it, and the temperature reaches the 3 degrees Kelvin temperature of its surroundings, wouldn't the black hole give up any further absorbed energy to its surroundings by radiation? The answer is the black hole can't give off any energy that falls into the black hole by radiation because the powerful gravitational attraction present at the event horizon of the black hole would prevent anything from escaping from inside the black hole, including the energy being absorbed.

The only radiation that a black hole can give off is Hawking radiation, an external radiation that could take millions and even billions of years to cause it to evaporate and vanish. The positive energy absorbed by the black hole is much greater than the small amount of energy given off by the Hawking radiation, because the positive energy available for absorption by the black hole is the huge amount of energy that permeates the universe. As soon as any positive energy is absorbed from space by the black hole more positive energy rushes in to replace the energy that was absorbed by it. This rapid absorption of positive energy occurs not only because there is so much positive energy available in space but also because of the strong attractive gravitational force present at the event horizon.

The net effect of this action is that the object will continue to absorb more positive energy than negative energy, resulting in a net increase of positive energy. As more and more positive energy is absorbed, it causes the internal temperature of the object to increase, resulting in an increase in the internal pressure of the black hole. This process continues until the internal pressure becomes strong enough to overcome the strong gravitational attraction of the object. In this state the object would again radiate light and energy. This rapid absorption of positive energy prevents the black hole from vanishing.

Thus, we see that if the black hole absorbs negative energy, as proposed by Hawking, the black hole would eventually vanish, creating the paradox of the missing information. However, long before the black hole could evaporate and vanish, it will absorb a considerable amount of the positive energy that was left over from the big bang, causing a rapid reversal of the process that created the black hole. In this state, the black hole could no longer exist because its strong gravitational attraction would be counteracted by the huge buildup of pressure generated within the black hole. Since the rapid absorption of positive energy prevents the black hole from vanishing, the paradox of the missing information doesn't occur.

Therefore, although there is no known experiment that proves the black hole theory wrong, it is difficult to understand, logically, how a black hole can exist for very long with its event horizon immersed in a sea of almost limitless energy.

# 22.
# Black Holes in The Two State Universe

Gravity in the two state universe is the attraction between the complements of mass and energy, not the attraction between masses. However, since the force of gravity derived from the two state universe theory gives essentially the same result as the force of gravity derived from Newton's equation of gravity, stars will form in essentially the same way in the two state universe as they do in the three dimensional universe. However, the force causing the gas molecules to be attracted to each other, in the two state universe, is caused by the attraction between the mass of one molecule and the energy of another molecule.

During the later stages in the life of a star, when the star loses all of its nuclear energy, the star collapses either to a white dwarf or a neutron star depending upon the size of the star that collapses.

If the star loses all of its energy during the collapse, then there wouldn't be any energy remaining in the star to create a gravitational attraction between the lifeless mass remaining in the star. The mass of the star, under these circumstances, would consist of pure mass. Such a star couldn't become a black hole, because, according to the two state universe theory of gravity, this star could only attract energy, it can't attract mass. And if the star could only attract energy, then such a star would be prevented from collapsing any further, because as the star attracts energy, it can only increase in size due to the increased internal pressure generated in the star as it continues to attract and absorb energy. Most of the energy attracted by the star would probably come from the massive amount of energy permeating the universe, that was left over from the big bang.

However, it is not logical to assume that a star would lose all of its energy during its collapse. It would be more logical to assume that the star would contain both mass and energy regardless of the type of star it collapses to.

For example, if the star collapses to a neutron star, the star would not only contain the binding energy that is normally contained in each of the neutrons, but it would also contain the huge amount of energy that is needed to keep the neutrons bound to each other.

Under these circumstances, the star will also have a considerable amount of gravity attracting the neutrons to each other. But regardless of how big the star was before it collapsed, even if it exceeded the Chandrasekhar limit, it could never collapse to a singularity or a point in space, because in order for a star to collapse to a point in space, it would need to generate a gravitational force that is infinite. This is because the volume of the neutron mass, as it collapses under the force of gravity, gets smaller and smaller. And as the volume of the mass decreases, the density of the mass increases. The density of an object being equal to the mass divided by the volume.

If the volume of the mass decreased to a point in space or a singularity, then the density of the star would increase to infinity, but before the density of the star can reach infinity, it would require an infinite force of gravity to accomplish this.

If we use the analogy that an object can't reach the speed of light because the mass of an object increases with the velocity, reaching an infinite mass at the speed of light. Then this infinite mass would require an infinite force to cause it to travel at the speed of light. However, the four dimensional time-space universe theory teaches that the force can only approach infinity it can never reach infinity, so the mass can never reach the speed of light. If from this four dimensional time-space universe theory analogy we could never obtain an infinite force needed to cause an object to travel at the speed of light, then we could never obtain an infinite force of gravity needed to collapse a star to a point in space, because the density of the mass contained in the collapsed star becomes infinite, at least for an instant, when the volume of the mass in the collapsing star becomes zero. We will examine this state of an infinite density at a point in space in more detail later.

Further, in order to create a singularity the mass being compressed by gravity would have to be made out of nothing, not matter. Since mass is matter and by definition mass occupies space, gravity wouldn't be able to collapse the mass to a zero volume, because no matter how much pressure is applied to the mass, the mass will offer more and more resistance to the force of gravity as the space it must occupy gets smaller and smaller. Eventually it will offer an infinite resistance to the pressure of gravity. And if the mass must occupy space then the space it occupies can't be reduced to zero. Therefore, the star could not collapse to a singularity.

Also, the collapsed star would have such a strong gravitational attraction at its event horizon that it will absorb, as we described in the last chapter, a considerable amount of the positive energy that permeates the universe. When the collapsed star absorbs this huge amount of energy, it will cause the temperature of the star to increase. And as the temperature of the star increases, it results in an increase in the internal pressure of the star.

When the star absorbs enough positive energy, the internal pressure will eventually become strong enough to overcome the strong gravitational attraction of the star. At this stage the star would again be able to radiate light and energy. And if during its collapse the star did manage to become a black hole, we could see that it wouldn't remain a black hole very long. Even if the collapsing star absorbs mass and energy from other stars, the same logic described above would apply.

In passing it is interesting to note that from the two state universe theory any pure or basic mass that exists outside the event horizon of the black hole would not be attracted to the black hole, because there wouldn't be any gravitational energy radiating outside the event horizon of the black hole. This is because all of the energy contained in the black hole would be strongly bound to and confined to the basic mass contained inside the event horizon of the black hole—nothing not even energy can escape from the black hole. The only way that basic mass can be attracted to the black hole is by becoming bound to any energy that exist outside the black hole, and when this energy is attracted to and absorbed by the black hole the mass that becomes bound to it would be carried along with the energy as it becomes gravitationally attracted to the black hole.

Complex mass could be attracted to the black hole because complex mass consists of basic mass and energy. The energy part of the complex mass radiates gravitationally out to the event horizon of the black hole. It is then attracted to and absorbed into the black hole, and when it is absorbed into the black hole it carries its associated complementary basic mass along with it.

Mainstream physics believes that when a star having a mass greater than the Chandrasekhar limit collapses, it will have a gravitational attraction that is so strong that this gravitational attraction would be strong enough to cause the mass and the energy contained in the star to be compressed to a point in space or a singularity. And in so doing, it will create a black hole.

Because the analysis described above appears to be in conflict with mainstream physics, we will examine this problem further and in more detail from the two state universe theory, using two different approaches. Both of these approaches are based on conclusions, laws and postulates derived from the two state universe theory.

We know from conclusion 26 of the two state universe theory that mass and energy are complements. They cannot exist in the same space at the same time. They can exist only if they are kept separated in time or space.

From conclusion 29 we know that gravity is the attraction between the complements of mass and energy. And from law 1 we know that complements cannot exist at the same point in space at the same time. They can exist at the same time only if they are kept separated in space. See conclusions 26 and 29 and law 1 in the appendix.

Since mass and energy exist at the same time, they must be space-separated complements. Therefore, they must be kept separated in space in order to exist as complements.

Using this information from the two state universe theory, we can now examine the problem from two different approaches.

In one of the approaches we will assume that the strong gravitational attraction will try to collapse only the mass contained in the star to a point in space. And if the gravitational attraction is strong enough to collapse the mass to a point in space, then only the energy would remain, and if only the energy remains, there would be no further gravitational attraction within the star, so the energy would escape from the star and be attracted to other mass in the universe.

In the second approach we will assume that the strong gravitational attraction will try to compress all the mass and energy contained in the star to a point in space, but could the gravitational attraction contained in the star become strong enough to accomplish this type of a collapse?

Let us examine the first approach. The strong gravitational attraction tries to collapse only the mass contained in the star to a point in space.

From this approach we see that as the strong gravitational attraction continues to compress the mass, the volume of the mass decreases and its density increases. As the volume of the mass decreases towards a zero point, the density will approach an infinite density. However, the gravitational force would have to become infinite to compress the mass to a zero volume with an infinite density. But can the force of gravity become infinite under these conditions?

We know from the two state universe theory, that gravity is the attraction between mass and energy and it is inversely proportional to the square of the distance between them. We also know that mass and energy are complements and they must be kept separated in space to exist.

In this first approach we assumed that only the mass is to be compressed to a point in space, these conditions means that the mass and the energy will always be separated by a finite amount. They can never have a zero separation between them because in order for this to occur the mass and the energy would have to come together at a point in space. This could never occur in this approach because we assumed that the energy was not to be compressed to zero with the mass.

From the inverse square law, if the mass and the energy are always separated by a finite amount and they never come together at a point in space, then the gravitational force will never become infinite. And if the gravitational attraction never becomes infinite, it will never be strong enough to collapse the mass of the star to a point.

Let us now examine the second approach, which is the more logical way that gravity reacts in a collapsing star. In the second approach we assume that gravity compresses both the mass and the energy to a point in space. Here again the gravitational attraction can't become infinite because the mass and the energy are space-separated complements. This means that there will always be a finite distance separating the mass and the energy, so the gravitational force never becomes infinite. Therefore, it never becomes strong enough to compress both the mass and the energy to a point in space.

The mass and energy can only have a zero separation between them when they are compressed to a point in space, but they can't be compressed to a point in space until the gravitational force becomes infinite. However, gravity can't become infinite before the mass and the energy are compressed to a point in space. An effect can't occur before the cause. An effect must occur after the cause; and since in this case it is the effect that is needed to create the cause, the effect, of the mass and the energy being compressed to a point in space, can't occur.

If, however, for some unknown reason the force of gravity does become great enough in a collapsing star to cause the mass and its associated energy to become so greatly compressed that the complements of mass and energy are forced to occupy the same point in space, then under these conditions, when the complements come together in the same point in space at the same time, they will cease to exist.

As more and more of the mass and the energy contained in the star are forced to occupy the same point in space and vanish, the star will lose more and more of its mass until the star eventually reaches a mass that is less than the Chandrasekhar limit. At that point the star will become a stable neutron star that doesn't have enough mass left for the effects of gravity to cause it to become a black hole.

Using the above logic, it is difficult to understand how black holes can exist. If black holes do exist, they can't exist for very long. Also, we have seen that a star can't collapse to a singularity or a point in space. This logic applies regardless of whether we use the three dimensional universe theory, the four dimensional time-space universe theory or the two state universe theory.

A paradox occurs, however, when the mass and the energy are compressed into a point in space. The paradox occurs because when the mass and the energy are compressed to a point in space, their density must increase to infinity. But when mass and energy occupies the same point in space, they will cease to exist, because they are space separated complements, and when space separated complements come together in the same point in space at the same time, they will cease to exist. They can only exist if they are kept separated in space. How then can we have mass and energy occupying a point in space in which their density is infinite and also zero?

Infinite density and zero density are complementary conditions. Complementary conditions can only be created or exist in the G factor. Refer to postulates 3 and 4 in the appendix. Postulate 3. states that the universe contains the G factor. And postulate 4. states that in the G factor, complements can exist at the same point in space at a fixed interval in time.

Therefore, we see that, when the mass and the energy are compressed to a point in space, their density will increase to infinity, but only at a fixed interval of time. This state immediately changes to the zero state when the mass and the energy vanishes. At this interval of time, the density goes to zero. Thus, the point in space could contain both of these states in the G factor, because a point in space is two state in nature. It can be in a state of infinite density and when it changes states it would then be in the state of zero density.

Mass always occupies space. It ceases to occupy space when it combines with its associated energy at a point in space. When it does both the mass and the energy, being space separated complements, will cease to exist.

From the above we see that when mass and energy are compressed to a point in space during an interval in time, their density and gravity increases to infinity, but since mass and energy are complements occupying the same point in space they immediately cease to exist and they vanish. This causes the point in space to change to the nothing state of no mass, no energy, no density and no gravity.

Thus we see that when mass and energy are compressed to a point in space, the point in space or the singularity could not remain in the infinite density and gravity state for more than a brief duration of time; and except for that brief duration of time, the singularity changes to and remains in the nothing state of zero density and zero gravity.

Therefore, we conclude that, although it is difficult to understand how black holes can exist, we can understand that black holes could possibly exist; but if they do exist they can't exist for very long. We also conclude that a star can't collapse to a singularity having an infinite density and infinite gravity.

# 23.
# The Unified Field Theory

For at least 150 years scientists tried to derive an equation that would unify the force of gravity and the electromagnetic forces in one unified field theory equation.

Michael Faraday believed that there was a relation between the force of gravity and the electromagnetic forces. He believed that this relationship could be demonstrated experimentally. In 1849 he wrote in his diary that this relationship between the force of gravity and the forces of electricity and magnetism should be capable of being demonstrated experimentally.

However, after many unsuccessful experiments, unable to experimentally show the relation between the three forces of nature, he suspended his experiments even though he strongly believed that there must be a relation between these three forces.

About one hundred years later, Albert Einstein tried mathematically to develop the unified field theory. He attempted to generalize his theory of relativity to include the relation between his theory of gravity and the electromagnetic theory. He worked unsuccessful for many years, until his death in 1955, trying to unify these three forces of nature.

The key to understanding the unified field theory depends upon our understanding of the force of attraction that exists between mass and energy. There is no question that such a force of attraction does exist between mass and energy. Therefore, it is hard to understand why the scientists, throughout the history of science, never derived or even tried to derive an equation that would describe this strong force of attraction between mass and energy. It is not clear if they ever realized that such a force existed.

They were able to derive an equation for the electrostatic forces that exist between the basic building blocks of mass—the positive particle and the negative particle. They were able to derive an equation for the magnetic forces that exist between the basic building blocks of energy—the north pole and the south pole of a magnet, and they also derived an equation that unified the electric and the magnetic forces; but they never derived an equation describing the force of attraction that exist between mass and energy. They never derived this equation because they probably never realized that such a force existed.

This failure of the scientists to derive an equation for the attraction between mass and energy probably occurred because the scientists always believed that mass and energy, being so strongly attracted to each other, was merely mass. They always considered complex mass to be just pure mass. They failed to keep the complements of mass and energy separated. This is another example of how the scientists failed to understand the theory of complements and the two state nature of the universe.

When Newton derived his theory of gravity he did exactly that. He considered the mass plus the energy of the apple, the moon, the earth, the sun and all bodies to be basic mass.

It is hard to understand how they could consider pure mass to be attracted to pure mass or pure energy to be attracted to pure energy. How could they separate the mass from the energy to test for these effects?

It is true that the positive and negative complements of pure mass attracts each other and the north pole and the south pole complements of pure energy attracts each other, but the same complements of mass, such as, two positive particles or two negative particles repels each other. The same is true for the same complements of energy—two north poles repel and two south poles repel.

Thus we see that the force of attraction between two pure masses is equal to the force of repulsion between the same two pure masses. These two equal and opposite forces results in the force between the two masses being equal to zero. This means that two pure masses neither attract nor repel each other, because the resultant force acting between them is zero. The same is true for the forces acting between pure energy.

In my book titled, the two state universe, I used Newton's theory of gravity to derive an equation for the force of attraction between two deuterium atoms, separated by a distance of one meter. I selected the one-meter distance because the force of gravity is directly proportional to the product of the two complex masses, and it is inversely proportional to the square of the distance between the two masses. By using one meter of separation, between the masses, the inverse square law effect would be equal to one, so the force of gravity under this condition would not be affected by the inverse square law, it would only be affected by the product of the two complex masses.

The deuterium atom consists of a nucleus of one proton and a neutron with an electron revolving around the nucleus. This means that the basic mass of the deuterium atom would consist of two protons and two electrons. The deuterium atom also contains basic energy. We represented the energy contained in the deuterium atom by two magnets. We can do this because energy can be converted from one form to another, so it doesn't matter if the energy in the atom is potential energy, kinetic energy or binding energy. The total combined energy contained in the atom would be converted to magnetic energy in which the total energy of the atom would be contained in the poles of the magnets. Because two magnets have four poles (two north poles and two south poles), the energy per pole would be equal to one quarter of the total energy contained in the atom.

Using Newton's equation for gravity, and substituting basic mass plus basic energy for each of the complex masses of deuterium, we find that when we multiply the two complex masses of basic mass plus basic energy times basic mass plus basic energy, we get a product having three terms. The three terms are basic mass times basic mass plus 2 times basic mass times basic energy plus basic energy times basic energy.

If we examine these three terms, we realize that the basic mass times the basic mass term would give us the resultant force that exist between the two protons and the two electrons that are contained in each of the deuterium atoms. Since there are an equal number of positive and negative particles contained in each atom, the attraction forces would equal the repulsion forces giving a resultant force of zero between the basic mass particles of the atoms. This means that the force acting between the two basic masses is equal to the electrostatic forces between the two atoms, which in this case is equal to zero. Thus the basic mass times the basic mass term is equal to zero.

Someone might ask how do we know that the basic mass times the basic mass term is equal only to the electrostatic forces acting between the mass particles?

We know from the two state universe theory that only complements have forces acting between them. The complement of basic mass in not another basic mass, the complement of basic mass is basic energy. There are no other complementary forces acting between the mass particles other than the forces that exist between the basic building blocks of the mass. These basic building blocks of mass, the proton and the electron, do have a complementary nature between them, which as we know from the two state universe theory, should have forces of attraction and repulsion between them. These forces between the basic building blocks of basic mass is the electrostatic forces.

Also if we divide the electrostatic force into the gravitational force between the basic mass of two deuterium atoms, we get an equation that shows that the force between the mass of the two deuterium atoms is equal to a constant times the electrostatic force between the two atoms, and even though there is a small constant associated with the electrostatic force, the resultant electrostatic force between the basic mass particles of the two deuterium atoms is equal to zero. Therefore, we see that there is no gravitational force of attraction between basic mass particles.

Further, it is believed that what we consider to be positive charge and negative charge is nothing more than the natural attraction and repulsion effects that occur among the basic complementary mass particles.

Looking at the basic energy times the basic energy term, we realize that this term is the magnetic forces between the two atoms. Because there are two north poles and two south poles in each of the atoms, the resultant force acting between these magnets would also be equal to zero, because the attraction forces are equal to the repulsion forces.

And like the effects of mass, the basic energy contained in each of the atoms does not have any forces of attraction or repulsion acting between the atoms except for the forces of attraction and repulsion resulting from the complementary effects between the basic building blocks of the energy—the north pole and the south pole of the energy. These forces are the magnetic forces acting between and among magnets.

There is no reason for us to believe that there are any other complementary forces acting between the energy contained in the two atoms except for the magnetic forces of attraction and repulsion already described between the magnetic complements of the north pole and the south pole. And like mass, we also believe that what is considered to be the north pole and the south pole of the energy is nothing more than the attraction and repulsion effects acting among these basic magnetic complementary building blocks of energy.

And if we divide the magnetic force into the gravitational force between the basic energy of the two deuterium atoms we get an equation that shows that the force is equal to a constant times the magnetic force acting between the energy of the deuterium atoms. And even though the constant associated with the magnetic force is small, the resultant magnetic force acting between the energy contained in the two deuterium atoms is equal to zero.

Thus we see that the resultant force acting between the energy contained in the two deuterium atoms is also equal to zero.

If the mass times the mass term is zero and the energy times the energy term is also equal to zero, then the gravitational force must be equal to only the mass times the energy term. This finding is in agreement with our assumption that gravity is the attraction between the mass and the energy. This mass times the energy term derived by multiplying complex mass times complex mass in this chapter is in exact agreement with the mass times the energy equation we derived in the gravity chapter, in which we assumed that gravity is the attraction between the complements of mass and energy.

In this chapter we derived the equation for the force of gravity that exists between two deuterium atoms. However, if we generalize the equation by using any two elements, we find that one element would have m protons and electrons and m north poles and south poles with the other element having n protons and electrons and n north poles and south poles.

Because there are an equal number of protons and electrons in one element and an equal number of protons and electrons in the second element, we find that the force of attraction between the particles is equal to the repulsion forces between the particles, giving a resultant force of zero acting between the masses of the two different elements.

The same is true between the magnetic poles of the elements. If there are an equal number of north and south poles in one element and an equal number of north and south poles in the second element, the force of attraction would be equal to the repulsion forces, giving a net force of zero acting between the energy of the two different elements.

The mass times the energy term results in a force of attraction between the mass and the energy that does not go to zero. This mass times the energy term is what causes the gravitational attraction between complex masses. It results because of the natural force of attraction that exists between the complements of mass and energy.

These results are amazing and unexpected because they not only confirm our findings that gravity is the attraction between mass and energy, but the results also shows that the force of gravity contains the three basic forces of nature (the electrostatic force, the magnetic force and the basic gravitational force of attraction between the elements) unified in this new gravitational equation. This new theory of gravity resulted because of the natural attraction that exists between the complements of mass and energy.

These results are also amazing because they not only confirm the existence of the unified field theory, but the result also shows that mass as well as energy has weight. This finding gives us some understanding of how the scientists believed, for many years, that the basic mass particles had a different weight when they were free and not bounded in an element from the weight they had when they were bounded in different elements.

The mass of the particles appeared to be different under different bound conditions, because when the particles were bound up in different atoms, the bound particles required different binding energy to keep them bounded in the nucleus of the different atoms. The binding energy required to keep the particles bounded in the nucleus depends upon the number of particles that are being bounded in the nucleus of the atom. However, since binding energy has weight, it would appear that the different particles had different weights depending upon the amount of binding energy that is needed to keep the particles bounded in the nucleus of the different atoms.

The unified field theory clearly shows that the weight of the basic particles does not change regardless of how they are bounded in the different atoms or even if they are not bounded in any of the elements. The same is also true about energy. Energy has weight because it has a gravitational force of attraction to the basic mass of the earth. It doesn't matter how the energy is distributed or associated with mass, because a unit of energy will always have the same unit of weight no matter how it is associated with mass.

It is this gravitational attraction between the mass and the energy that caused the scientists to believe that the atomic energy was being created by the loss of the mass of the particles that were bounded in the nucleus of the atom. They believed this because the weight of the uranium atom was greater than the weight of the particles that resulted after the explosion. The energy given off during the explosion had weight when it was bounded in the nucleus of the atom, so it is only natural that the weight of the particles given off during the explosion would appear to weigh less than when they were bounded in the nucleus of the uranium atom. Thus we see that the unified field theory eluded scientists for at least 150 years because the scientists believed that gravity was the attraction between masses. They failed to understand complements and the two state nature of the universe, and they also failed to understand the force of attraction that exists between mass and energy.

From the results described in this chapter, we were able to make the following conclusions which were first described in my book titled, the two state universe:

30. The unified field theory unifies the three basic forces of nature in the new gravitational equation derived from the two state universe theory.

31. The unified field theory supports the theory that gravity is the force of attraction between the complements of basic mass and energy.

32. The unified field theory shows that gravity is a complex force of the sum of the following forces between these complex masses: electromagnetic, electrostatic and gravitational.

The unified field theory described in this chapter supports the gravitational theory derived in the chapter on gravity, and it also supports the two state universe theory that flashed into my mind, like a bolt of lightening, many years ago, as my brother and I were driving back to Connecticut, from our trip to his house in Forest City, Pennsylvania, in 1983.

# APPENDIX

## Conclusions

1. The universe is made up of infinite numbers of points in space.
2. Each point contains the existence or the nonexistence of something.
3. The universe is dynamic; it is continually changing and is in motion.
4. The universe contains things that exist as well as the non-existence of things.
5. Each point in space, at any one time, could contain the existence or nonexistence of something.
6. Each point in space is two state in function.
6a. Definition of complements: Each of two parts or conditions are complements if the truth of one means the other is false. If it is true that at one point in space something exists, then the nonexistence of it at the same point at the same time would be false. If it is true that at one point there is the nonexistence of something, then the existence of something at the same point at the same time would be false. Therefore, each point can be in one or the other complementary conditions but cannot be in both complementary conditions at the same time. Thus, each part of the complement is a state; since there are two parts to a complement, and since the point can be in either of these two complementary condition, then each point in space must be capable of being in one or the other state. Because of this, we must conclude that each point in space is two state in function.
7. The universe functions as a two state universe and is a two state universe.

8. Energy is important for motions of the universe. Without it there could not be a two state universe.
9. Mass is essential for the existence of a physically solid two state universe.
10. The two state universe does indeed depend upon energy and mass for its existence and function.
11. Everything that exists, exists as a complement in our two state universe.
12. Time is important in the two state universe only in that it is man's perception of increments of the ever changing two state universe. It is neither energy nor mass, neither physical nor solid.
13. Space is nothing or a void. However, it is just as real and meaningful as something since it is the complement of not space, which is something.
14. Opposites of likes attract.
15. Energy repels itself when emanating from a source and traveling in the same direction.
16. Complex mass attracts complex mass.
17. Light is a complement consisting of a wave and a particle complementary part.
18. The wave and the particle complement of light are space-separated complements.
19. It is the two state nature of the universe that gives it its jellylike nature required to transmit energy through space and yet offer no resistance to a material object such as planets revolving in their orbits.
20. Light travels through the aether space at a velocity:

$$c = \frac{dx}{dt} = \frac{\text{distance between points in space}}{\text{time it takes the points to change states}}$$

21. Light travels through the aether space at a velocity that is constant and independent of the velocity of the source that produced it.
22. Although light travels through the aether space at a velocity that is constant and independent of the velocity of the source that produced it, the velocity of light as measured by an observer would be affected by the velocity of the observer.

23. Once a ray of light has a particular path in the aether, it continues in that direction; it changes only if it is acted on by another object or force of energy so as to reflect it or interfere with that path.
24. The velocity does not have any affect on the contraction of an object. That is, an object does not contract in the direction of the velocity.
25. Time does not decrease with an increase in velocity; it does not change with velocity.
26. Mass and energy are complements. They cannot exist in the same space at the same time. They can exist only if kept separated in time or space.
27. Mass in motion is a mass energy complement where the energy and the mass are kept separated in space.
28. Basic mass does not increase with velocity but rather the energy of the mass increases with velocity, and this energy is a complement of the basic mass.
29. Gravity is the attraction between the complements of mass and energy.
30. The unified field theory unifies the three basic forces of nature in the gravitational equation derived from the two state universe theory.
31. The unified field theory supports the theory that gravity is the force of attraction between the complements of basic mass and energy.
32. The unified field theory shows that gravity is a complex force of attraction between complex masses consisting of the sum of the following forces between these complex masses: electromagnetic, electrostatic, and the gravitational force derived from the two state universe theory.

# Laws

Law 1. Complements cannot exist at the same point in space at the same time. They can only exist at a point in space at different times or they can exist at the same time only if separated in space.

Law 2. Different parts of different complements can occupy the same point in space at the same time, if, and only if, the different parts are neither complementary in function nor both physically solid.

Law 3. Everything that exists in the universe exists as a complement. Complements can exist only if they are kept separated in what man perceives as time or space.

Law 4. Something that exists cannot be in two different places at the same time; however, it can be in two different places at different times.

Law 5. like parts, of a real pair complement repel; unlike parts attract.

# Postulates

Postulate 1. A point in space is two state in function.

Postulate 2. A point in space is not divisible.

Postulate 3. The universe contains a G factor.

Postulate 4. In the G factor, complements can exist at the same point in space at a fixed interval in time.

Postulate 5. Creation can occur only in the form of complements.

Postulate 6. Creation is possible in the G factor.

# Part
# 2

## New Conclusions

33. Based on the results of the laser experiment, using the earth as a moving platform, we conclude that there is an absolute reference frame—the aether.
34. The laser experiment shows that you can use optical and electromagnetic principles to determine the velocity of your reference frame.
35. The laser experiment also teaches that not only is there an absolute reference frame—the aether, but there is no absolute inertial frame—a reference frame where an object can travel in a straight line with a constant velocity.
36. If an electron is prevented from changing its orbit, it will not radiate energy.
37. An electron is prevented from changing its orbit, as it revolves around the nucleus of an atom, because the electron generates an electromagnetic field in such a direction that if the electron starts to radiate energy it will began to be pulled into the nucleus. When the electron moves towards the nucleus, the induced magnetic field generates a force on the electron that causes the electron to increase its velocity. An increased velocity increases the centrifugal force causing the electron to move away from the nucleus. It is these balanced forces that prevents the electron from changing its orbit as it revolves around the nucleus of an atom, in a steady state condition, in an allowed orbit.
38. An allowed orbit is an orbit that an electron can revolve around in because there is no magnetic energy present in

this orbit, and the forces acting on the electron are in balance. These are the conditions needed to prevent an accelerating electron from changing orbits and radiating energy.

39. An accelerating charged particle does not always radiate energy.

40. Mass doesn't change to energy in an atom bomb, because there are the same number of basic particles after the explosion as there were before the explosion, and the basic mass particles neither increase nor decrease in mass. The basic mass particles merely rearrange themselves into different elements that contain a total binding energy that is less than the binding energy that was contained in the uranium atom. Also, the total energy contained in the atom bomb before the explosion is the same as the energy given off during the explosion plus the binding energy contained in the particles given off after the explosion.

41. As a charged particle revolves faster and faster in a magnetic field, the radius of curvature of its circular path increases with the velocity. When the charged particle reaches the speed of light, the radius of curvature of the charged particle's path increases to infinity. This means that the charged particle is traveling in a straight line unaffected by the magnetic field.

42. Electromagnetic radiation is a complement of a charged particle and its generated magnetic field traveling at the speed of light. This electromagnetic complement traveling at the speed of light, through the aether, travels in a straight line unaffected by the magnetic field.

43. An accelerating charged particle generates a magnetic field which creates an electric field in the form of at least two complementary charged particles. The charged particles travel in opposite directions to each other along the envelope of the electric field, which is oriented at 90 degrees to the orientation of the magnetic field. When the two oppositely charged particles meet half way around the electric field orbit they annihilate each other creating a bundle of magnetic energy. These bundles of energy are created twice during each cycle of the electromagnetic wave, as it travels in the aether. Therefore, energy travels

in a discontinuous manner through the aether in the form of bundles of energy.
44. It is believed that these bundles of energy are the quanta of energy described by Planck.
45. These quanta of energy, or photons, would appear to be massless because the particles creating them are annihilated as the bundles of energy are created.
46. In an electromagnetic wave, we find that the wave travels at the speed of light in the aether, but the particles generated in the electric field travel around the envelope of the electric field with a velocity that is 1.57 times the speed of light.
47. If we divide the distance between the points in space (the Planck length) by the time the points in space takes to change states (the Planck time) we get a velocity equal to the speed of light in the aether.
48. There is a limiting frequency that an electromagnetic wave can have as it travels in the aether. This frequency is equal to $1.85 \times 10^{43}$ hertz.
49. There is a maximum photon energy that can travel through the aether space. This energy is equal to$12.21 \times 10^{9}$ joules.
50. Energy can travel through the aether in a continuous manner in the form of kinetic energy.
51. It is possible for a particle to travel through the aether faster than the speed of light because the mass of the particle doesn't increase with velocity.
52. The kinetic energy of a particle can travel through the aether with a speed greater than the speed of light, because if the particle of the kinetic energy travels faster than the speed of light, the points in space would not have enough time to change states from the nothing state to the energy state. In this state the particle travels unaffected through the aether space because the aether remains in the nothing state, and the kinetic energy, being so highly attracted to the particle, because of their complementary nature, would be carried along with the particle through the aether space. Under these conditions, the energy doesn't have to wave anything to travel through the aether because it would be

carried through the aether by the strong attraction of the fast moving particle.

53. A particle can and does travel through space faster than the speed of light, and there is no reason known from the two state universe theory why the particle can't travel at any speed.

54. The event horizon of any black hole is immersed in a sea of almost limitless energy, the energy left over from the big bang. In this state the black hole could easily absorb enough energy to replace the energy that was lost during its creation. In effect the process that created the black hole is reversed, because the pressure generated by the huge amount of energy absorbed by the black hole would counteract the strong gravitational attraction of the black hole, allowing anything—including light and energy—to again escape.

55. It is difficult to understand how a black hole can exist, for more than a short duration of time, with its event horizon immersed in a sea of almost limitless energy.

56. Mass by definition must occupy space. If mass must occupy space then the space it occupies can't be reduced to zero. Mass would cease to exist if it was compressed to a point in space. Therefore, a star can't collapse to a singularity of infinite density and infinity gravity.

57. Gravity can never compress mass and energy to a point in space, because according to the two state universe theory gravity is the attraction between mass and energy. To compress mass and energy to a point in space, gravity would have to become infinite, but gravity can't become infinite before the mass and the energy are compressed to a point in space. Therefore, the force of gravity never becomes infinite. An effect can't occur before the cause.

58. If mass and energy are compressed to a point in space they will cease to exist because they are complements. Complements cannot exist unless they are kept separated in either time or space.

59. When mass and energy are compressed to a point in space in the G factor, their density will increase to infinity, but only for a fixed interval of time. This state immediately changes to the zero state when the mass and the energy

vanishes because they are complements. Therefore, a star can't collapse to a singularity having an infinite density and an infinite gravity.

60. The key to understanding the unified field theory depends upon our understanding of the force of attraction that exist between mass and energy.

61. There is no gravitational force of attraction between basic mass particles if there is an equal number of positive and negative particles.

62. It is believed that what we consider to be positive charge and negative charge is nothing more than the natural complementary attraction and repulsion effects that occur among the basic mass particles.

63. There is no gravitational force of attraction between basic energy particles if there is an equal number of north and south poles.

64. Like mass, we also believe that what is considered to be the north pole and the south pole of a magnet is nothing more than the natural complementary attraction and repulsion effects acting between the magnetic poles of magnets.

65. If the basic mass times the basic mass terms between two complex masses is equal to zero and the basic energy times the basic energy terms is equal to zero, then gravity must be equal to the force of attraction between the basic mass and the basic energy terms.

66. Mass and energy have weight.

67. Energy has weight because it has a gravitational force of attraction to the basic mass of the earth.

68. Because binding energy has weight, it would appear that the basic particles contained in the different elements have different weights because the weight of the binding energy needed to bind the nucleus of the different elements would be different for each of the elements.

69. The unified field theory shows that the weight of the basic particles do not change regardless of how they are bounded in the different atoms or even if they are not bounded in any of the elements. It is the weight of the binding energy of the different elements that changes, not

the weight of the basic mass particles that are bounded in the different elements.

70. In an atom bomb, the weight of the binding energy of the uranium atom is greater than the combined weight of the binding energy of the particles given off after the explosion. It is this difference of the binding energy weights that caused the scientists to believe that the mass of the particles contained in the uranium atom decreased and changed to energy when the atom bomb exploded.

# BIBLIOGRAPHY


1. Asimov, Isaac, *The Subatomic Monster*, New American Library, A Mentor Book, New York, 1968

2. Augros, Robert M. and Stanciu, George N., *The New Story of Science*, Bantam Books, New York, 1986

3. Boslough, John, *Steven Hawking's Universe*, William Morrow and Company, Inc., New York, 1985

4. Clark, David H., *Superstar*, McGraw-Hill Book Company, New York, 1984

5. Cork, James M., *Radioactivity and Nuclear Physics*, D. Van Nostrand Co. Inc., Toronto-New York-London, 1947

6. Dull, Charles E., Metcalf, H. Clark and Williams, John E., *Modern Physics*, Holt Rinehart and Winston, Inc., New York, 1960

7. Eddington, Sir Arthur, *Space Time and Gravitation*, Harper & Brothers, New York, 1959

8. Einstein, Albert and Infield, Leopold, *The Evolution of Physics*, Simon and Schuster, New York, 1938, 1966

9. Einstein, Albert, *Relativity*, Crown Publishers, Inc., New York, 1966

10. Ennes, Harold E., *Boolean Algebra for Computer Logic,* Howard W. Sams & Co. Inc., Indianapolis, Indiana, 1978

11. Feinberg, Gerald, *Solid Clues*, A Touchstone Book, Simon and Schuster, New York, 1985

12. Ferri, Timothy, *The Red Limit*, Quill, New York, 1983

13. Field, George B. and Chaisson, Eric J., *The Invisible Universe*, Birkhouser, Boston, 1985

14. Gamow, George, *The Great Physicists from Galileo to Einstein*, Dover Publications, Inc., New York, 1988

15. Gardner, Martin, *The Relativity Explosion*, Vantage Books, New York, 1962, 1976

16. Gribbin, John, *Spacewarps,* Delacorte Press, New York, 1983

17. Halliday, David and Resnick, Robert, *Physics for Students of Science and Engineering*, John Wiley & Sons, Inc., New York-London, 1960, 1962
18. Hritzsch, Harold, *The Creation of Matter*, Basic Books, Inc., New York, 1984
19. Knight, David C., *Isaac Newton*, Franklin Watts, Inc., New York, 1961
20. Lorentz, H. A., Einstein, A., Minkowski, H., and Weyl, H., Translated by Perrett, W. and Jeffery, G. B., *The Principle of Relativity*, Dover Publications, Inc., 1923
21. March, Robert H., *Physics For Poets*, Contemporary Books, Inc., Chicago, 1970, 1978
22. Molinaro, Angelo A., *The Two State Universe*, Branden Publishing Company, Boston, 1988
23. Pais, Abrahm, *Inward Bound*, Claredon Press-Oxford-Oxford University Press, New York, 1986
24. Reichenbach, Hans, *From Copernicus To Einstein*, Dover Publications, Inc., New York, 1970
25. Rothman, Tony, *Instant Physics*, Fawcett Columbine, New York, 1995
26. Rusk, Rogers D., *Introduction to College Physics*, Appleton-Century-Crofts, Inc., New York, 1954, 1960
27. Russel, Bertrand, *The ABC of Relativity*, New American Library, New York, 1958, 1969
28. Sears, Francis Weston, *Principle of Physics*, Addison-Wesley press, Inc., Canbridge, MA., 1947
29. Semat, Henry, *Atomic Physics*, Rinehart & Company, Inc., New York, 1939, 1946
30. Shortley, George and Williams, Dudley, *Elements of Physics*, Prentice-Hall, Inc., Englewood Cliffs, New Jersey, 1971
31. Strathern, Paul, *The Big Idea-Newton and Gravity,* Anchor Books Doubleday, New York-London-Toronto-Sidney-Aukland, 1997
32. Taffel, Alexander, *Visualized Physics*, Oxford Book Company, New York, 1940
33. Taylor, Edwin F., and Wheeler, John Archibald, *Spacetime Physics*, W. H. Freeman and Company, San Francisco and London, 1963, 1966

34. Weber, Robert L., White, Marsh W. and Manning, Kenneth V., *College Technical Physics*, McGraw-Hill Book Company, Inc., New York, and London, 1947

# ABOUT THE AUTHOR

Angelo A. Molinaro, author of the book titled, *The Two State Universe*, attended the Bullard Havens Technical School and the Bridgeport Engineering Institute in Bridgeport, Connecticut. Later, he attended the Massachusetts Institute of Technology where he received his Bachelor of Science degree in Electrical Engineering. Upon graduating from MIT, he was employed by United Technologies Corporation, Sikorsky Aircraft Division, where he worked on a number of projects including Special Instrumentation, Dynamics, Acoustics, Sonar, Electrostatics and Navigation. He participated in the helicopter qualification tests and blind flying test at Patuxent River Naval Air Station in Maryland, sonar test at the Naval Air Station in Key West, Florida, and Operation Plumbbob (atomic bomb tests) in Nevada. He holds a number of diplomas and degrees, including a Master of Science and a Doctor of Science degree in Electrical Engineering from Ohio Christian College and a Doctor of Philosophy degree in Theoretical Physics from Northwestern International Univerisity. In addition, he holds patents on two inventions; he also holds an Amateur Extra Class Radio Operator License and is a member of The American Radio Relay League. Mr. Molinaro is presently living in Monroe, Connecticut.

Made in the USA
Monee, IL
04 November 2026

67383330R00098